T0323618

Safety and Health Competence

A Guide for Cultures of Prevention

Taylor and Francis Series in The Interface of Safety and Security

George Boustras
European University Cyprus, Nicosia, Cyprus

Published Titles

Safety Management in Small and Medium Sized Enterprises (SMEs)
George Boustras, Frank W. Guldenmund

Safety and Health Competence
A Guide for Cultures of Prevention
Ulrike Bollmann and George Boustras

For more information about this series, please visit: https://www.crc press.com/The-Interface-of-Safety-and-Security/book-series/CRCINT SAFSEC

Safety and Health
Competence

A Guide for Cultures of Prevention

Edited by
Ulrike Bollmann and George Boustras

CRC Press
Taylor & Francis Group
Boca Raton London New York

CRC Press is an imprint of the
Taylor & Francis Group, an **informa** business

CRC Press
Taylor & Francis Group
6000 Broken Sound Parkway NW, Suite 300
Boca Raton, FL 33487-2742

© 2021 by Taylor & Francis Group, LLC

CRC Press is an imprint of Taylor & Francis Group, an Informa business
No claim to original U.S. Government works

Printed on acid-free paper

International Standard Book Number-13: 978-1-138-61173-3 (Hardback)
International Standard Book Number-13: 978-0-367-50760-2 (Paperback)

Visit the Taylor & Francis Web site at
http://www.taylorandfrancis.com

and the CRC Press Web site at
http://www.crcpress.com

*Umuntu ngumuntu ngabantu: A person is
a person through other people.*

To our African friends

Contents

SECTION II   Case Studies

Foreword

There is a song by Van Morrison that I particularly like – "Enlightenment (I don't know what it is)". You should listen to it; it is really enlightening. Well, at least for me.

I sometimes have this same feeling about culture – I don't know what it is. Although I have been thinking about it for more than 20 years, not all the time, of course, the concept of culture is often slipping through my mental fingers. I do not seem to get a real grip on it. The late Barry Turner has been quoted as saying – defining culture is like nailing jelly to the wall. It is, it really is.

Why is culture so hard to define? And why is culture still so popular amongst scientists and practitioners? There are several answers to these questions. To begin with, culture "emerges" (this is really the best word I can come up with) amongst a group of people when they interact with each other, within a particular context, which makes this context and what they do within there, meaningful (in the sense that they continuously assign meaning to and try to make sense of what is happening). Here, culture is about shared understanding, about shared meanings. Such understandings might be written down, or tacit, or both, but they are not arbitrary. This is the dynamic part of culture. However, there is also a more static part to it: culture at large. For instance, the context in which people work and assign meaning to the things they do and say to each other is embedded in a larger, much less dynamic culture. This might be an organizational culture, but also that culture is embedded in a larger culture. Here, culture is about more fundamental issues, such as, how you relate to your boss and how (s)he relates to you (i.e., power, and how it is expressed and effected). About what is polite and impolite (i.e., rituals), which and how emotions can be shown, and so on. This is the culture you can assume being present when you encounter a group of people from the same part of the world and which helps them to get through everyday life without too much hassle. These are the rules of the never-ending game called Life.

Apparently, the more one says about culture, the more complex it becomes. The culture jelly by now starts to drip from the nail and I have not even started yet.

Another way of nailing down culture is by looking for metaphors that capture its assumed or intended meaning. A useful metaphor enables us to explore culture by drawing comparisons between the metaphor's attributes and the (possible) attributes of what is meant by culture; that is, culture as imagined by those who want to apply it (the manager, the scientist, the practitioner). I think we need several metaphors here to encapsulate most of culture's intended usages. Here are a few to consider.

The invisible hand. This metaphor might especially appeal to managers as it is their invisible hand, guiding their followers along the rocky road of safety. If they might get tempted to go astray, the invisible hand puts them on the right track again. The invisible hand does not relent and it is omnipresent and omniscient. This might also be the metaphor many managers have in mind when they talk about safety culture.

The social glue. This metaphor emphasizes the (assumed) binding power of culture. This might be what people mean by – "We have a strong culture". It stresses uniformity, loyalty, peer pressure, to be together on the same page where safety is concerned. It also invokes caring for each other, helping each other, looking out for each other, but also turning a blind eye, because harmony should be preserved and conflicts avoided.

The contest. Within this metaphor, culture is both a game and something that can be good or bad, or something in between. It invokes ladders, tugs of war, a bandstand, and frontrunners and laggards. This metaphor is for winners or umpires who like to compare and judge. Or for showboats who want to show off how well they are doing regarding safety. The contest also implies a game, which is played according to the rules. If you don't play to the rules, you're disqualified. With safety, the game is not straightforward, and rules will need to bend. How much slack do we allow in playing the safety game? When will you be disqualified? And what is the purpose of the game? "Zero accidents" is hardly a result to show on the scoreboard. Zero points usually means – game over. And what is the nature of the game? Are we opponents, rivals, competing for the same prize? Or are we a team, with shared responsibilities and complementing competencies? And who do we play the game for, who's our audience, and how can we keep them satisfied?

The (in)convenient truth. Truths can be wanted or unwanted, convenient or inconvenient. The convenient truth serves us well, the inconvenient truth does not. Why would we be interested in a convenient truth? With convenient truths, things remain the way they are. We can stay put. Until an inconvenient truth comes along. It might be that we think we're quite safe. We might believe that we have everything under control and that we can take a break, finally. We feel relieved. We become complacent. And then an inconvenient truth comes along. We have an accident. We're not as safe as we thought we are. What will we do? Will we take convenient measures or inconvenient ones?

A convenient truth can also be something else. We can treat safety culture as a label to obscure responsibilities. When the culture is to blame, everybody is to blame and therefore no one is to blame. Safety culture can be a great smokescreen to cover up wrongs and abuses.

The book that lies before you is about cultures of prevention and about the competences required to achieve this. This is something people should be good at, as we are evolution's champions. We are survivors. But things also have become very complicated these days. Much common sense has been replaced by common rules. And common rules cannot cover everything. We need to improvise; we need to learn. But we can only learn from our mistakes, so we need to make mistakes and we need an environment where making mistakes is possible, wanted, encouraged. Paradoxically, a culture of prevention should enable this.

I encourage you to look for mistakes and metaphors in this book and beyond. To embark on the learning journey with inconvenient truths and challenged by convenient ones. Always stay close to this path and do not merely trust the invisible hand but also trust your common sense. You'll need it.

Frank Guldenmund

Work and workplaces vary a lot around the world and so is the focus and competence of and on safety and health. In this book authors from all five continents give their ideas and their own experience to achieve a culture of prevention. While in African and Central Asian countries the socio-economic conditions are desolate, most industrialized countries around the world are in a transformation guided by globalization, digitalization and immigration. Authors address all these aspects, both theoretically as from a very practical point of view.

Paul Swuste

Acknowledgments

The editors would like to thank Soon Ang, prof. and chair of the Division of Strategy, Management, and Organization at the Nanyang Business School, Singapore; Monika Eigenstetter, prof. for industrial and organizational psychology at the University of Applied Sciences – Hochschule Niederrhein, Krefeld, Germany; Frank W. Guldenmund, PhD, ass. prof. at the Safety Science & Security Group at Delft University of Technology, the Netherlands; Hiltraut Paridon, prof. and director of studies at Bachelor's and Master's degree in medical education at the SRH University for Health, Gera, Germany; Sabine Sommer, director of science at the Federal Institute for Occupational Safety and Health (BAuA), Berlin, Germany; Paul Swuste, PhD, ass. prof. em. at the Delft University of Technology, the Netherlands and all authors who have contributed to this book with their expertise and advice.

Special thanks go to Andrew Corrigan for the careful editing of the manuscript as well as to Kati van der Seylberg for her creative support with the book cover and Alexandra Walk for the natural and quick help with the graphics.

Editors

Ulrike Bollmann is Head of International Cooperation at the Institute for Work and Health (IAG) of the German Social Accident Insurance (DGUV). After studying education and philosophy, Ulrike finalized her academic studies with a thesis on the historical-logical development of the modern concept of knowledge. She worked as a researcher at the Regional Institute for Schools and Further Education of North Rhine-Westphalia (1991–1994), and as a project manager at the West German Trade Council (1994–1999). In 1999, she started working at the IAG of the DGUV. From 2002 to 2004, she was employed as a national detached expert at the European Agency for Safety and Health at Work (EU-OSHA) in Spain. Ulrike has been the founder and coordinator of the European Network Education and Training in Occupational Safety and Health (ENETOSH) since 2005. The network has 97 members from 38 countries and holds numerous MoUs, for example with the Robert W. Campbell Award, IOHA and OSHAfrica. In 2009, 2011, 2013, and 2016 she conducted the International Strategy Conference on Safety and Health at Work in Dresden in cooperation with WHO, ILO, ISSA, EU-OSHA, IALI and others. She played a major role in organizing the XX World Congress on Safety and Health at Work – Global Forum for Prevention in 2014 in Frankfurt and held a symposium at the XXI World Congress in Singapore in 2017. She was responsible for an empirical study on OSH education and training and conducted a joint research project with the Korean Occupational Safety and Health Administration (KOSHA) on the leading indicator "trust" for a culture of prevention. Ulrike is a member of the Editorial Board of Safety Science, a member of the Scientific Committee on Education and Competency Development, the Training and Advisory Council of OSHAfrica, and member of the Advisory Board of the Erasmus+ project RiskMan.

George Boustras is Professor in Risk Assessment at European University Cyprus, Dean of the Ioannis Gregoriou School of Business Administration and Director of the Centre of Risk and Decision Sciences (CERIDES).

George is a PhD in Probabilistic Fire Risk Assessment from CFES at Kingston University London, he was Honorary Research Fellow at CPSE at Imperial College London (2003–2005), and KTP Research Fellow at FSEG at the University of Greenwich (2009).

He sits at the Management Committee of Secure Societies – Protecting Freedom and Security of Europe and its citizens of "HORIZON 2020". George has been invited to present his and CERIDES' work at a number of organisations (e.g. Imperial College, JRC Ispra, University of Malaga, University of Dalian etc).

He was appointed by the Ministerial Council of the Republic of Cyprus to Head the Special Task Force that overlooked the modernization of the Fire Services. He was hired by World Bank to contribute to the modernisation of licensing services provided by the Fire Service of the Hellenic Republic. The President of the Republic of Cyprus appointed him, as Vice President in the Energy Strategy Council. He consulted the Ministry of Defence of the Republic of Cyprus in the Risk Assessment of Unexploded Ordnance as part of Gas Exploration.

George is Editor-in-Chief of Safety Science (Elsevier) and Member of the Editorial Board of Fire Technology (Springer), the International Journal of Emergency Management and International Journal of Critical Infrastructure (both Inderscience). He (co-)supervises 7 PhD students.

List of Contributors

Ulrich Birner
Siemens AG
Munich, Germany

Ulrike Bollmann
Institute for Work and Health (IAG)
    German Social Accident Insurance
    (DGUV)
Dresden, Germany

Helen Bound
Research Centre for Work and Learning
    (CWL)
Institute for Adult Learning (IAL)
Singapore University of Social Science
Singapore

George Boustras
Ioannis Gregoriou School of Business
    Administration
Centre of Risk and Decision Sciences
    (CERIDES)
European University of Cyprus
Nicosia, Cyprus

Florian Brendebach
University of Koblenz-Landau
Koblenz, Germany

Jason Edwards
Queensland University of Technology
Brisbane, Australia

Hanan Mohamed Fathy Elnagdy
University of Dibrugarh
Dibrugarh, India

Emily J. Haas
National Institute for Occupational
    Safety and Health (NIOSH)
Pittsburgh, Philadelphia

Dick Hoeneveld
Delft University of Technology
Safety and Security Science section
Delft, the Netherlands

Heinz Hundeloh
Statutory Accident Insurance for
    the public sector in North Rhine-
    Westphalia (UK NRW)
Düsseldorf, Germany

Toran Law
Occupational Safety & Health Council
Hong Kong SAR, China

Just Mields
German Social Accident Insurance
    Institution for the Energy, Textile,
    Electrical and Media Products
    Sectors (BG ETEM)
Köln, Germany

Dingani Moyo
Baines Occupational Health Services
Harare, Zimbabwe

Peter Paulus
Center for Applied Sciences of Health
    (CASH)
Leuphana University
Lüneburg, Germany

Rüdiger Reitz
Institute for Work and Health (IAG)
German Social Accident Insurance
    (DGUV)
Dresden, Germany

Markus Schöbel
University of Basel
Basel, Switzerland

Martin Schröder
German Social Accident Insurance
  Institution for the Energy, Textile,
  Electrical and Media products
  sectors (BG ETEM)
Köln, Germany

Steve Tsang
Occupational Safety & Health Council
Hong Kong SAR, China

Lawrence Waterman
Park Health & Safety Partnership
  LLP
London, UK

Kristine Yap
Research Centre for Work and Learning
  (CWL)
Institute for Adult Learning (IAL)
Singapore University of Social Sciences
Singapore

Bonnie Yau
Occupational Safety & Health Council
Hong Kong SAR, China

Patrick L. Yorio
National Institute for Occupational
  Safety and Health (NIOSH)
Pittsburgh, Philadelphia

1 Introduction

Ulrike Bollmann and George Boustras

CONTENTS

1.1 INTRODUCTION

Work and learning are currently undergoing an unprecedented transformation – never before has change been so broad, so fast, and above all so disruptive at the same time. Formal and non-formal qualifications and degrees are replaced by individual competence acquisition, competence development by organizations, and more recently, the design of learning processes in cross-organizational structures such as networks. In this transformation, informal learning becomes more and more important, not least because of digital and social media.

This book brings different perspectives to bear on the question of which competences are needed for the transformation of work in the face of globalization, digitalization, and migration. Which concept of competence is appropriate in the face of rapid change and the digitalization of all life and work domains? What is the impact of globalization, digitalization, and migration on competences related to occupational safety and health? Which concept of competence promotes a humane design of work?

The book is divided into a more theory-oriented basic part and a more practical part. In the basic part, authors representing different disciplines and different approaches explore the issue of competence in occupational safety and health for cultures of prevention. The range of approaches includes interpretative, constructivist, and cognitivist approaches, as well as approaches based on systems theory. In the practical part, the authors present case studies with an emphasis on competence development among a specific target group. Here, the spectrum ranges from supervisors on construction sites to workers on construction sites and tea plantations, safety professionals, and school heads.

The book is intended for occupational safety and health experts, human resource developers, educational institutions, training designers, teachers, and trainers. It has been conceived in a conscious effort to cross the boundaries between occupational safety and health and education.

The authors and co-authors in this book come from 11 countries and all 5 continents: Africa (Egypt, Zimbabwe), Asia (Hong Kong, Singapore), Americas (United

States), Australia, and Europe (Cyprus, Germany, Switzerland, the Netherlands, the UK).

This first chapter continues with an introduction to the concept of competence (1.2) and the concept of a culture of prevention (1.3), followed by summaries of all chapters (1.4).

1.2   EVOLUTION OF THE CONCEPT OF COMPETENCE

The term competence as it is used today originates from motivational psychology (White 1959) and linguistic theory (Chomsky 1965). In the 1970s, the term went on to become popular in education and training. In the final decade of the 20th century, the term competence replaced the term qualification. This "competence-oriented turn" was also a response to changes in overall conditions at the workplace (Sprafke 2016, 69). "Competence" has become ubiquitous since the year 2000 in the wake of the PISA studies (Programme for International Student Assessment), when debates were dominated by the question of whether competences could be measured, and if so how. The ubiquitous and increasingly narrow use of the term culminated in the OECD competence strategy, which described competences as "the global currency of the 21st century" (OECD 2012).

Doubts about the predominant view of competence served as the inspiration for this book. The prevailing instrumental view of competence is largely a static one: competencies are generally understood as consisting of integrated pieces of knowledge, skills, and personal attitudes. Accordingly, current attempts to address the transformation of work rely on repeatedly revised competency catalogs, combined with blind faith in the standardization of competencies. Competence development is measured exclusively in terms of learning outcomes, which means that competence assessments or measurements are disconnected from the learning process (cf. Bollmann 2018).

A universally accepted concept of competence does not exist. The minimum consensus is the idea of competence as "what people are able to do" (Mulder 2017, 12). In the Anglo-Saxon world, the matter is further complicated by having to distinguish between the terms *competence*, *competency* (plural: *competencies*), and *competent* (see Bound and Yap (Chapter 2) and Schöbel (Chapter 5) in this book).

It may help to distinguish between different "waves" or views on competence. Martin Mulder (2017) offers the following distinction:

(1) behavior-oriented view (functional approach; e.g., a curriculum featuring specific, small tasks in which task performance is rewarded: see, for example, Elnagdy, Chapter 10 in this book);

(2) socio-constructivist view (integrated approach; knowledge, skills, and attitudes need to be addressed together, personal development should be based on knowledge and skills formation: see, e.g., the formation of a professional identity in Waterman's contribution to this book in Chapter 8); and

(3) the relationship between context and conceptualization (the meaning of competence and competencies is specific to each context: see, e.g., Bound and Yap (Chapter 2) and Schöbel (Chapter 5) in this book).

This is not a clear-cut distinction, and the contributions in this book draw on blends of all three approaches.

Furthermore, Mulder distinguishes between two fundamentally different ways of operationalizing competence:

(1) behavior-oriented generic competence: the emphasis here is on generic content- and context-free competency statements; as an example of this, see Schöbel's reference to the "great eight competencies" of Dave Bartram (Bartram 2005) in Chapter 5 in this book; and

(2) task-oriented specific competence: good examples of this can be found in the field of medicine (e.g., Cate ten 2017); for the tasks of safety professionals, see the design of a competency framework by Moyo (Chapter 7), and the development of a training model by Reitz and Schröder, Chapter 11 in this book.

In addition, this book introduces an alternative understanding of the concept of competence: the capability approach (Nussbaum and Sen 1993; Terzi 2014; see Bound and Yap, Chapter 2 in this book, with reference to Holland et al. 1998). According to the capability approach, occupational safety, health, and wellbeing not only depend on people's functionings (their "beings" and "doings") but also on their capabilities. A key issue in this context is the freedom to choose a specific way of life, in this case a specific workplace, for example (cf. Elnagdy, Chapter 10 in this book) but also the freedom to learn and work in a self-determined manner (cf. Brendebach (Chapter 4) but also Yau, Law, and Tsang (Chapter 9) in this book). In other words, a capability is a kind of freedom: namely, the substantive freedom to pursue different functioning combinations (Sen 1999, 95). This freedom is not one that is inherent in a person; it rather describes freedoms or opportunities that arise from the combination of personal abilities and the specific political, social, and economic context (Nussbaum 2011, 29). Whether it is learning or work, the key point is to create opportunities that enable people to create and lead lives they can value and have reason to value (Otto and Ziegler 2012, 40).

Capabilities cannot be described in competency catalogs; they refer to the task of providing each individual with a work and learning environment that gives them access to safety and health or enables them to choose good work and safe and healthy actions (based on Nussbaum 2012). The capability approach contains the (yet unrealized) potential to arrive at a more precise understanding of what defines a humane work environment under the conditions of globalization, digitalization, and migration (Sustainable Development Goal 8, "Decent work and economic growth"). As a general principle, the capability approach draws attention to the fact that material resources are an issue that must be considered seriously and in a systematic manner without pitting them against the cultural dimension (Otto and Ziegler 2012, 41).

1.3 THE CULTURAL TURN IN PREVENTION

The cultural turn in prevention in the occupational context notes a shift of prevention away from averting and avoiding risks towards proactive safety and health management and empowerment. The focus of attention in companies, administrations, and educational institutions is now on the informal side of organizational learning (Schein 2004; Guldenmund 2015). Prevention becomes a socially grounded process of meaning-making. Safety, health, and wellbeing are made meaningful through the interactions and communications of people in a specific context: their family, their peer group, their group of coworkers, or their relationship to their supervisor.

From this perspective, a culture of prevention is not only rooted in a shared collective understanding of safety, health, and wellbeing; rather, it also always encompasses a diverse set of realities, which is why the title of this book refers to cultures of prevention in the plural (Berger and Luckmann 1973; Blazsin and Guldenmund 2015; Guldenmund 2015; Bollmann et al. 2020).

According to the three "ages of safety" identified by Andrew Hale as early as 1998 (technology, human factors, and safety management), occupational prevention has now entered the age of (interconnected) human relationships:

> The culture of prevention concept sees occupational safety and health expanding or changing its perspective from a technical or risk-oriented view to the issue of designing relationships: How do we want to work and live in the future? Those work contexts in companies, public administrations, and educational institutions that support the involvement of people and social relationships in work processes in addition to promoting safety and health are also likely to be more successful in managing dynamic technological and demographic challenges. (Bollmann et al. 2020)

Instead of merely reacting to the "technological drivers" – complexity, acceleration, uncertainty, volatility – a culture of prevention calls for promoting the "human drivers" – diversity, meaningfulness, freedom, flexibility (Dobiey 2018; Bollmann and Eickholt 2019).

A joint project of the Institute for Work and Health of the German Social Accident Insurance (DGUV) and the Korea Occupational Safety and Health Agency (KOSHA) developed indicators for predicting and promoting safety and health in a workplace setting. Researchers in this project identified and categorized 17 leading indicators under the 5 domains of communication, leadership, participation, learning process, and trust (Bollmann et al. 2020). Broken down into these categories, the competences required for cultures of prevention to emerge may be summarized as follows for this book:

In terms of *communication*, competence development is not limited to promoting instrumental skills; it rather means looking at a specific workplace to identify the qualities of a positive culture that foster good communications and respectful relationships between individuals and teams (Waterman, Chapter 8 in this book). *Leadership* must be reconceptualized: as the outcome of a dialogical process involving the active participation of employees (Mields and Birner, Chapter 3 in this book); as interaction (Schöbel, Chapter 5 in this book); as an attribute and capability of an organization (Mields and Birner); as safety leadership (Schöbel) and salutogenic leadership (Paulus and Hundeloh, Chapter 12 in this book). Ideally, full engagement and *participation* foster a sense of belonging together, a "collective competence" that is essential for a culture of prevention to emerge (Bound and Yap, Chapter 2 in this book). Participation means that employees are experts in their own competences (Brendebach, Chapter 4 in this book). Furthermore, the transformation of work is accompanied by new forms of teaching and learning: *Learning* takes place in a self-directed manner (Brendebach); cross-cultural interactions not only have to be learned, they also have to be maintained (Yorio et al., Chapter 6 in this book); virtual learning facilitates expanded learning experiences in the area of safety and health (Yau et al., Chapter 9 in this book); learning based on concrete tasks promotes competence development (Reitz and Schröder, Chapter 11 in this book). Interpersonal *trust* (in the capabilities of employees) but also basic trust in the world (and its institutions and systems) are the foundation of a culture of prevention in the occupational context (Schöbel, Brendebach, Paulus, and Hundeloh in this book).

1.4   THE CONTRIBUTIONS IN THIS BOOK

This section provides summaries of the contents of each chapter.

Chapter 2, Helen Bound, head of the Center for Work and Learning (CWL) at the Institute for Adult Learning (IAL) at Singapore University of Social Sciences, opens this book with a keynote chapter on "Reconceptualizing 'Developing Competence at Work' to a Journey of Being and Becoming", which is based on an ethnographic study by Kristine Yap. Bound follows the agency approach, drawing on work by Dorothy Holland.

Safety and health competence is more than merely the ability to act in a safe and healthy manner; it refers to a contextual and collective process of meaning-making. In this context, Bound and Yap speak of "agency" as the sum of all possible actions a person has at their disposal. Being and becoming, negotiating and re-negotiating professional identities are active social processes that require agency to be able to shape the transformation of work and to change cultures of prevention. Professional identities and agency are mutually dependent.

By adopting this perspective, Bound and Yap reject the traditional understanding of competence as a set of static attributes of knowledge, skills, and attitudes. They follow a holistic approach that does not view individuals in isolation but always embedded in their biographical and social context. In contrast to behaviorist, cognitive, and constructivist understandings of competence, their one is an interpretative approach:

> From the interpretative perspective, competence is about workers (individually and collectively) developing understandings of the work, as opposed to an attribute-based only perspective that assumes that the mind and the world are separate entities. This perspective allows for a development perspective of individuals and collectives as they interact with their world.

By contrast, traditional competency-based training (CBT) is based on a dualism of mind and body, theory and practice, and individual and collective. That is why CBT has little to contribute to the evolution of a culture of prevention, as the focus is on the acquisition of knowledge and skills by the individual rather than on the social construction and creation of knowledge. There is systematic evidence supporting Bound's and Yap's analysis in scientific reviews showing that competence training per se has few long-term effects on occupational safety and health if the competence training fails to reflect and support the culture and the context in which the competences are to be practiced later (Robson et al. 2010; van Dijk et al. 2015; Schulte et al. 2017; see also Swuste and van Dijk 2019).

Accordingly, Bound and Yap emphasize the importance of the workplace as a "learning space" for the development of occupational safety and health practices in the sense of changing cultures of prevention. The workplace offers a wealth of learning experiences in which individual perceptions are being continuously negotiated and shaped by competing discourses in the physical and social context of work. This is essential for a shared understanding of occupational safety and health practices.

Bound and Yap illustrate the agency approach using a number of concrete occupational safety examples taken from Yap's qualitative research on Singapore's petrochemical industry. The examples show how exchanges and communication may have positive effects on the evolution of safe and healthy practices in the workplace. A key prerequisite for this to happen is the existence of an institutional framework allowing for workers' full engagement and participation, thereby facilitating collective learning processes. Ideally, this fosters a sense of belonging together, a "collective competence" that is essential for a culture of prevention to emerge. The emergence of cultures of prevention requires agency for existing practices to change. To Bound and Yap, the transformation and evolution of cultures, whether with respect to occupational safety practices or other issues, represent a process of transformation far more complex than the traditional, functional conception of competence.

Chapter 3, "Culture of Prevention and Digital Change: Five Theses on Work Design", is devoted to the transformation of work, which is largely fueled by digitalization.

In this chapter, Just Mields, an occupational and organizational psychologist at the German Social Accident Insurance Institution for the energy, textile, electrical, and media products sectors (BG ETEM), and Ulrich Birner, an industrial psychologist and head of psychosocial health and wellbeing at Siemens AG, discuss five theses on the effects of the digital transformation on the nature of work and the individual and organizational competences needed to address these effects. Arguing from a systemic perspective, the authors propose five theses. Thesis 1 calls for an expansion of existing work system models. Thesis 2 describes work-related opportunities and risks associated with the digital transformation as areas of conflict. Thesis 3 confirms the positive effect of risk assessments on the design of future work environments. Thesis 4 discusses the importance of leadership as a function of the organization, and thesis 5 underscores the relevance of systematically developing a culture of prevention in an environment of a continuous dynamic of change.

Mields and Birner define digital transformation as a disruptive, fast-paced, and dynamic process of change fueled by digital technologies and affecting all levels of society, including companies. The authors highlight the fact that it is not just individual factors but the entire work system that is affected by the digital transformation. When designing work systems, therefore, it is important, first, to be aware of the permanent need for adaptation and the effects of the system's dynamic self-organization.

Second, leadership needs to be reconceptualized as the outcome of a dialogic process involving the active participation of employees. Employers must ensure a participatory space as a means to achieve the best possible work design. Third, they need to keep in mind that corporate culture has a crucial impact on the way the technical and social elements of the system interact with each other.

Thesis 2 is used to demonstrate the current polarization of work as a result of the digital transformation. Using the example of a Siemens project on the effects of the digital transformation on safety, health, and the environment, the authors show how the influence of the digital transformation is perceived as a conflicting mix between opportunities and risks when it comes to occupational safety and health.

Thesis 3 emphasizes the importance of holistic risk assessment. In work systems undergoing digital transformation, we see a growing share of assessments devoted to mental stress and strain. Furthermore, it is important to identify assessment and design procedures that take account of new, emerging types of stress caused by digital technologies and that compensate for the lack of established scientific knowledge by instituting participatory processes of risk assessment. Risk assessments must include all workplaces where employees are engaged in work systems transformed by digital technology. Data and information about the work process must be included in real time in the risk assessment in an effort to dynamize the risk assessment procedures themselves in the process.

In their fourth thesis, Mields and Birner call for a new concept of leadership. From the authors' systemic point of view, leadership is no longer understood as the attribute or capability of a person but rather as the attribute and capability of an organization. In a world of work transformed by digital technologies, leadership is needed when reflecting on internal and external relationship management as well as on decision-making patterns in the face of fast-paced development.

In their fifth and final thesis, Mields and Birner directly address the dimension of culture. How can organizations foster the emergence of a culture of prevention in the context of digital transformation and as a way to shape the effects of that transformation?

Chapter 4, Florian Brendebach is a research associate at the Institute of Pedagogy at the University of Koblenz-Landau, Germany. His contribution to this book, "Competence Management: Between Command and Control, Self-Organization, and Agility", is an in-depth exploration of the aspect of cultural dynamics, enhanced by the concept of agility. He argues from a systemic-constructivist perspective.

Brendebach sees competence as a disposition for self-organization and hence as a potential source of autonomous and creative action. He defines competence management as an organization's efforts to influence this disposition in a systematic manner. The changing nature of work in an increasingly complex, dynamic, and unpredictable world serves as the backdrop to the author's argument. As competences are acquired in a self-organized manner but never in isolation, the organizational level is the focus of attention in this chapter. Brendebach contrasts a top-down concept of competence management with the idea of agile competence management, elaborating on what both of these concepts mean in terms of occupational safety and health.

The crucial element in the author's understanding of competence is the distinction between other-directedness and self-directedness. From his systemic-constructivist perspective, learning and competence acquisition are always self-organized processes based on individual structures. What makes the difference, however, is whether learning goals are defined in a self-directed or in an other-directed manner. Self-directed learning may facilitate a strong fit between the learner's internal structures and the learning goals selected. As a consequence, competences are acquired between the two poles of instruction and facilitation.

Drawing on Knut Illeris, Brendebach distinguishes four types of learning: cumulative, assimilative, accommodative, and transformative learning. In a four-field scheme, these types of learning are linked to the poles of other- and self-directedness, revealing distinct contexts of competence acquisition. An employer providing top-down safety instructions, for example, would constitute a case of other-directed assimilation, involving a clearly defined qualification requirement. Self-directed assimilation, by contrast, requires employers to trust their employees as independent actors in charge of their own learning.

In view of the digitalization of all life and work domains, Brendebach contrasts a "command-and-control mindset" to an "agile mindset", which follows two basic principles: (1) the needs of the customer are the paramount concern (cf. Mields and Birner, Chapter 3 in this book); and (2) decisions are made where the relevant competence is located. According to Brendebach, this means that although it may still be possible or indeed necessary for strategic decisions to be made at the center of an organization, their operational implementation should be carried out by comparatively small, interdisciplinary, and self-organized teams working where the organization meets its environment. The focus thus shifts from top-down management to the facilitation of self-organization.

Brendebach suggests three ways of introducing agility to competence management: (1) participation – employees are experts in their own competence; (2) dynamization of processes – reducing hierarchies may enable employees to experience a higher degree of freedom; and (3) self-organization – agile, interdisciplinary teams authorized to make independent decisions work to ensure a credible implementation of the organization's vision and mission.

In an agile organization, trust in employee's capacities for self-organized learning plays a major role. The transformation towards an agile organization must be understood as a developmental process that needs to be initiated from the top down. In taking this first step, decision-makers may benefit from coaching or an assessment of the organization's existing culture.

Chapter 5, "Managing Competencies of Safety Leaders: Some Promises and Shortcomings", describes the development of a competency model for safety leaders. The author, Markus Schöbel, is an organizational psychologist, researcher, and lecturer at the Department of Psychology at the University of Basel, Switzerland.

His arguments are based on existing studies on safety culture in organizations and the impact of leadership on safety. Schöbel discusses the potential for identifying and developing leadership competencies that help to improve work safety. Drawing on the "competency modeling" technique, he looks into the potential of that technique for systematically enhancing safety leadership behavior and safety culture. In doing so, he breaks new theoretical ground.

Schöbel's chapter is divided into two major parts. In the first part, he introduces the "competency modeling" technique and goes on to develop his competency model for safety leadership, which is grounded in empirical evidence. In the second part, he discusses the pros and cons of managing leadership competencies in the area of safety, describing the challenges and future research needs with regard to competency modeling for work safety.

The technique of competency modeling aims to match individual competencies and organizational strategies and objectives. Unlike existing competency models, which were either developed in relation to specific work contexts and work roles or to describe more general competencies of safety experts without focusing on leadership positions, Schöbel's model is a generic competency model for safety leadership. In developing that model, he builds on existing competence frameworks, drawing especially on Campbell's concept of leadership performance (as opposed to management performance, see Waterman Chapter 8 in this book) and the six subfactors contained therein, which relate to what leaders do: (1) consideration, support, person-centeredness; (2) initiating structure, guiding, directing; (3) goal emphasis; (4) empowerment, facilitation; (5) training, coaching; and (6) serving as a model.

Schöbel's generic competency model comprises 18 concrete performance-focused competencies of safety leadership, of which three are linked to each of Campbell's six subfactors, respectively.

In the second part of his contribution, Schöbel discusses the pros and cons of competency modeling. The advantage of having a precise profile of a safety leader, for example, is associated with the disadvantage that the competencies described are unlikely to work in all kinds of situations or may not be shown by one and the same leader. Competency models are more than a mere addition of competencies, and the effectiveness of a leader is not evident from the sum of their competencies.

According to Schöbel, substantial difficulties may arise when monitoring and evaluating safety leadership performance. Evaluating leadership competence on the grounds of accident statistics, for example, disregards systemic approaches to organizational safety and may even have counterproductive effects. That is why the development of leadership competence in the field of safety should always be embedded in broader scheme of organizational development. What is more, leadership behavior is always an interaction; what matters is the inter-individual combination of competences, not just the individual competencies.

Chapter 6, "Cultural Intelligence: A Construct to Improve Occupational Safety and Health in the Face of Globalization and Worker Mobility Across National Borders" focuses on cross-cultural competence. The main author, Patrick L. Yorio, works at the National Personal Protective Technology Laboratory of NIOSH, Pittsburgh, Philadelphia. His team of co-authors represents multiple nationalities and academic disciplines.

The authors examine the challenges for occupational safety and health (OSH) managers arising from the increasing cultural diversity in the workforce. In particular, rising worker mobility across national borders poses cultural challenges for OSH managers who may be accustomed to workers from a specific cultural background. To deal with these challenges, the authors highlight the concept of cultural intelligence (CQ). By applying this concept to occupational safety and health, Patrick L. Yorio and his co-authors break new theoretical ground.

The chapter provides background on migration-related labor movement, introduces CQ as an important competence for OSH managers, and describes examples of how OSH practices may differ across cultural contexts.

The authors define cultural intelligence as "an individual's capability to effectively manage themselves and others in cross-cultural situations and environments". To explore this further, they discuss models by Soon Ang and colleagues and by David C. Thomas. They cite a wide range of empirical studies proving that specific CQ dimensions predict outcomes such as cultural judgment and decision-making; enhanced psychological wellbeing, greater satisfaction, and commitment while working in foreign national cultures; task and group performance; and effective leadership.

The authors follow a four-factor model that describes CQ as a form of human intelligence and as a holistic ability. The model consists of the following four, empirically verifiable dimensions: metacognitive, cognitive (knowledge), motivational, and behavioral. By including motivation, Yorio et al. reject the idea of restricting CQ to the dimensions of cognition/knowledge, abilities, and metacognition, thereby highlighting the importance of the motivational component. Persons who are not interested in cross-cultural interactions, or who are unwilling to learn or maintain such interactions, are unlikely to be good candidates when it comes to establishing cross-cultural initiatives at the organizational level. This point is particularly important with a view to occupational safety and health.

In their application of CQ to occupational safety and health, the authors make additional reference to studies by Hofstede and three of his valued-based cultural dimensions, which have been empirically demonstrated to be at work in organizations: uncertainty avoidance, power distance, and future orientation. When applying CQ to occupational safety and health, it is important to keep in mind how people socialized in different national cultures react to different directives, procedures, programs, and leadership styles. Using examples of normative OSH strategies assigned to the selected dimensions by Hofstede, Yorio et al. show how cultural motivation and knowledge, cross-cultural behavior, and metacognitive dimensions of cultural intelligence may be applied to the context of occupational safety and health in cross-cultural situations in order to increase the effectiveness of managerial choices.

The authors see a need for further research and development with regard to the development of CQ forms specific to occupational safety and health and in the empirical analysis of the predictive validity for successful OSH management across cross-cultural settings.

Furthermore, the authors recommend an analysis of the potential strengths and weaknesses of different OSH strategies in cross-cultural settings that goes beyond existing cross-cultural comparisons of OSH practices.

Chapter 7, "Competencies in Safety and Health That Meet the African Complexity and How to Measure Them", was authored by Dingani Moyo from Zimbabwe. An occupational safety and health consultant and occupational medicine specialist, Moyo teaches at the University of the Witwatersrand in South Africa and at Midlands State University in Zimbabwe. He is the chairman of the Scientific Committee on Education and Competency Development of OSHAfrica, a pan-African occupational safety and health network.

With his contribution to this book, Moyo offers a first conceptual foundation for a pan-African competency framework for OSH practitioners and professionals.

The chapter is dedicated to generic competencies in occupational safety and health. Moyo points out that these competencies are indeed relevant for the African context. When developing his competency framework, he is guided by the capability framework of the International Network of Safety and Health Practitioner Organizations (INSHPO). The competency framework developed by INSHPO describes activities, skills, and expertise that are essential for OSH practice.

In the first part, Moyo begins by giving an overview of the basic OSH conditions in Africa; in the second part, he outlines the competency framework for OSH practitioners and OSH professionals in Africa; and in the third part, he presents a small selection of tools for the measurement of competencies.

According to Moyo, one key factor characterizing the occupational safety and health situation in Africa is the lack of access and the inferior quality of OSH services. He argues that improving the competencies of OSH practitioners and OSH professionals providing such services is one of the most important steps towards improving the quality of and access to OSH services. The key lever to systematically improve the working conditions at African companies is the introduction of occupational safety and health management systems (OSHMS).

Moyo describes the basic conditions of OSH and OSH training on the African continent as follows:

- On the African continent, the majority of workers are employed in the agricultural and mining sectors. The degree of mechanization is limited in both sectors, with most of the labor performed manually. This means that workers are exposed to diverse and substantial hazards, including exposure to pesticides, ergonomic hazards, and injuries.
- Africa is characterized by a huge diversity of traditions and cultures. There is one principle, however, that is shared by all African cultures – a principle expressed in the following proverb: "A person is a person through other persons".
- OSH practice in Africa is in a desolate condition. There is a lack of adequate training in OSH throughout Africa; the lack of well-trained OSH practitioners and OSH professionals creates a gap often closed with non-qualified staff; there is a shortage of occupational physicians, occupational hygienists, and work ergonomists, which further exacerbates the situation of OSH practitioners and OSH professionals.
- In addition, there is a growing informal sector that draws in vulnerable groups of the population such as children, older people, women, and pregnant women at random.

In the second part of his chapter, Moyo begins by addressing the core activity domains of OSH practitioners and OSH professionals in Africa, which he identifies as: communication, management and leadership, change management, promoting a culture of prevention, auditing, ethics, procurement and contracting, total quality management, and measurement. What he describes here are the practical demands on the activities of OSH practitioners and OSH professionals on the African continent, for example, clear communication in the face of low literacy rates, a good ability in dealing with people over and above the OSH technical knowledge, and a good skill set in managing OSH in diverse cultures in Africa with its five regions and diverse languages.

Aside from the absence of occupational safety and health laws and their less than satisfactory enforcement, corruption and abuse of authority is a major problem in many African countries. Under these circumstances, Moyo insists, OSH practice is in need of practitioners and experts capable of conducting themselves in line with professional and ethical standards.

The following chapters present five case studies, with each case study focused on one special target group.

Chapter 8, "Management and Leadership at Supervisor Level: The Black Hat Program", is about the competence development of direct supervisors in the construction industry. Lawrence Waterman shares insights about his work as Head of Health and Safety for the Olympic Delivery Authority for the London 2012 Olympic Games. After completing the first construction program in the history of the Olympic Games without a work-related fatality, Waterman was awarded the Order of the British Empire (OBE) by HM The Queen.

The chapter offers a description of the background to the construction program for the London 2012 Olympic Games, the development of a leadership program for managers and supervisors, and processes of cultural change initiated by the construction program.

The construction program was funded by the British government and commissioned and monitored by the Olympic Delivery Authority (ODA). The construction work was performed by a set of main contractors in a competitive setting. The ODA specifications contained a Health and Safety Standard with six priority themes: equality and diversity; employment and skills; design and accessibility; sustainability and legacy; defining how we work; and security as a strategic decision in face of terrorism.

Waterman describes the difficulties encountered on a large construction site. Frequent staff turnover meant that maintaining a safety and health culture, a feeling of "this is the way we work around here", with reliable, repeated behavior patterns was potentially harder to achieve than in more stable working environments with a relatively static workforce. Another challenge was the special characteristics of the construction workforce: a low level of education among builders and where workers are predominantly male, older, and reserved when it comes to health issues. Migrant work is another factor that does not help improve health care; "false self-employment" and fears of "blacklisting" rather impede the development of an engaged health and safety culture.

Against this background, the ODA created a program that covered all the elements of a comprehensive approach to occupational safety and health: planning, procurement, contracts, leadership program, and reporting. By defining the program's safety and health objectives and detailing some of the most important methods to be applied, the ODA established a framework documented in the ODA "Health and Safety Standard" and included in the construction contracts with the entire supply chain. Waterman believes that the ODA approach fostered more respect for the workers on the site and encouraged a much higher degree of engagement in the development of on-site work practices, not only in the monitoring of these practices.

Furthermore, a leadership program was developed that consisted of two levels. The first level was created by getting the executives and project managers of the major supply chain companies to meet once a month with ODA representatives and their supply partners as the "Safety, Health, and Environment Leadership Team" (SHELT). The second level was a training program for on-site supervisors, which was established for two reasons: (1) the leadership experienced by workers on the construction site is not the leadership of managers but that of their direct supervisors, the "effective day-to-day leaders of construction work"; and (2) direct supervisors serve as an important link between management and a constantly fluctuating workforce.

Waterman explicitly points out that this training program was not limited to improving supervisors' competencies and promoting the development of their instrumental skills; rather, it was about identifying the qualities of a positive site culture, the conditions that foster good communication and respectful relationships between individuals and teams. The outcome of these observations on "felt leadership [sic]" for site workers was, on the one hand, a program to raise supervisors' awareness of their authority, responsibility, and seniority. These qualities were made visible across all projects by requiring all supervisors to wear black hard hats, which over time became known as black hats. Moreover, supervisors were invited to program-wide meetings and events, the "Black Hat Conventions", which served as a forum for supervisors from various projects and teams to meet and discuss their work and the challenges related to health and safety. Waterman is convinced "that if we treated supervisors as leaders, it would encourage them to become leaders. (…) Generally, the conventions acted as a mechanism for generating pride in and respect for the role".

The ODA training program to develop supervisors as leaders consisted of the following elements: understanding their own role in risk management; developing communication skills; using listening skills, for instance, to give workers at the daily activity briefing a real opportunity to bring up problems and exchange all relevant information; and recognizing and questioning unacceptable behaviors and work methods.

According to Waterman, the training program took a broader approach compared with traditional, instrumental training to developing skills to oversee the work day-to-day and to intervene to achieve compliance with rules. The program changed the perception of safety and health being about following rules to it being an expression and proof that each job was being done in the right way.

The ODA training scheme became Common Standard 38 in the UK construction sector and was recognized as a new general industry standard by the UK Construction Group, a cross-sector association.

According to Waterman, the Black Hat Program pioneered a reassessment of where leadership is really located and how it can be developed. Even though the engagement of senior management is a key factor in all units of a company, "Vision Zero" can only be achieved if the entire workforce is engaged – and the London 2012 Black Hat Program has shown that supervisors play a key role in this.

Chapter 9, "Workers in a Virtual Work Environment: An Immersive Safety Learning Experience", Bonnie Yau and her team present a case study on virtual training for the construction sector. Bonnie Yau is executive director of the Occupational Safety and Health Council (OSHC) in Hong Kong SAR, China. The virtual training program is geared towards regular workers and trainees.

The authors provide a step-by-step explanation of why and how they selected the virtual mode of learning for safety training in the construction sector. The main driver in this decision was the search for an effective form of safety and health learning in times of rapid digitalization.

Based on studies on the effectiveness of training methods in the area of safety and health, the authors opted for a setting that is not only marked by a strong element of activation but is also one in which the degree of interactivity corresponds to the degree of hazard involved in a given situation.

The authors draw on experiential learning theory (ELT), based on David Kolb's four-stage learning cycle, as the theoretical foundation for the use of learner activation methods in the safety and health training.

As traditional forms of experiential learning are not only resource intensive but may also involve hazards of their own, the authors looked for an alternative method to physically simulate the real-world environment for various training tasks. The OSHC team found this kind of method in a virtual environment-based training system using a three-dimensional (3D) interface. The core of this method is using virtual reality to give users a sense of presence as a way of promoting their cognitive and spatial learning: "By applying VR technology for experiential learning, workers are able to completely immerse themselves in a virtual environment which mimics a realistic work environment for learning and training under different preset scenarios".

In response to the vast number of fatal accidents involving falls from a height in the Hong Kong construction sector, the OSHC developed the "Cave Automatic Virtual Environment" (imseCAVE) and in 2018 established the "OSH Immersive Experience Hall" in the OSHC Academy. The first scenario developed for imseCAVE was the "working at height safety" scenario.

The evaluation of the virtual training program revealed high and significant rates of participant satisfaction. Nearly all participants said the training served to increase their awareness of the hazards involved in working at height, which leads the authors to expect that such greater awareness may also translate into safer practices when working at height.

Yau and her team highlight the following imseCAVE features:

(1) Trainees may be given the freedom to make their own decisions without causing negative effects in reality.
(2) The virtual work environment is dynamic and may hence help improve risk awareness on the construction site.

(3) Based on trainees' knowledge gaps and learning needs, trainers may create different training scenarios, meaning that virtual training programs offer a high degree of flexibility in terms of adapting or personalizing training scenarios.

With regard to the evaluation of the virtual training program, the authors recommend collecting information on accident incidence in addition to surveying participant satisfaction, observing trainees in their work process, and reviewing the quality of preventive measures.

Chapter 10, "People-Oriented Teaching Intervention for Tea Plantation Workers in Assam: A Teaching Intervention Study", Hanan Elnagdy presents a case study on training Indian plantation workers with regard to occupational safety and health. A chemist by training, Elnagdy works for the Ministry of Manpower and Immigration in Cairo, Egypt and has a master's degree in green chemistry and OSH. She conducted this study in cooperation with Dibrugarh University in Assam, India.

In the first section of her presentation, Elnagdy offers in-depth insights into tea production in Assam and into the living and working conditions on the tea plantations and factories. In a second section, she discusses the design, implementation, and evaluation of an occupational safety and health training intervention performed on the tea plantations.

The socio-economic conditions in Assam's tea industry are desolate. According to Elnagdy, low wages, low education, unskilled workers, gender inequality, and the poor relationship between workers and management are common work-related problems on the tea plantations. The poor health situation (low standard of living, lack of sanitary facilities, low-quality food) is exacerbated by tea workers' unhealthy lifestyle (keeping domestic animals inside the house, consuming saltwater and alcohol). Caught in a vicious cycle of poverty and poor education, looking for different work is simply not an option for many plantation workers. The relationship between workers and management is not a good one, whereas the supervisors play a positive role, as Elnagdy points out.

With respect to occupational safety and health in Assam's tea industry, Elnagdy sees workers exposed to all kinds of hazards (chemical hazards, biological hazards, physical hazards, including agronomic hazards, solar radiation and fire exposure, and psychological hazards). Overall, workers' lack of awareness concerning occupational safety and health, as well as a lack of OSH skills, increases the risk of their involvement in accidents and work-related illnesses on the tea plantations.

In the second section, Elnagdy describes the design, implementation, and evaluation of the training program she conducted in Assam. She defines training as "a common type of intervention that focuses on enhancing the level of knowledge of specific groups of people in order to motivate their behavioral change". To achieve that goal, she believes it is essential to select a method that best suits the specific target group. That is why she made field visits prior to the intervention to analyze the actual situation on the plantations. She visited both the tea bushes and the tea factories, as well as the workers' communities, the so-called "workers' lines".

The teaching intervention had the following objectives: (1) to increase workers' knowledge about occupational safety and health; (2) to enhance personal behaviors to avoid risks in the workplace; and (3) to improve relations between workers and management. To accomplish these objectives, the teaching intervention consisted of two measures. The first involved tips for good personal hygiene and the proper use of personal protective equipment; the second was teaching workers about good working postures in different positions.

Her method was based on colorful, clear images and posters depicting safe and unsafe working postures, for example. Moreover, bad posture, for instance when carrying baskets and nets, were simulated and compared with the good ones. The intervention featured games like play ball, an exercise with raising red and green cards, and others. The idea was to motivate workers by involving them in the activities, like sharing in games, simulating the postures, pointing at photos, and the like.

The goal of the evaluation, which was conducted by asking direct questions through the interpreter and by playing games, was to find out whether tea workers' knowledge of occupational safety and health had increased, whether they had changed their behavior, and whether the relations between workers and management had improved.

In conclusion, Elnagdy reflects on three crucial lessons learned: (1) OSH is a powerful and unique language that is wordless, simple, direct, easy to understand, attractive, and effective; (2) bring the service to the door of the target group; and (3) a teaching intervention is a remarkable approach in itself. To Elnagdy, the goal of the intervention was workers' development. She wanted to help make safety and health concerns an integral part of tea workers' lives by empowering them to express themselves and showing them suitable and simple OSH solutions to be adopted easily with little effort and in line with their budget.

Chapter 11, "New Competence of Safety Professionals: A Comprehensive Approach", explores the recent redesign of a training program for safety professionals in Germany. The author, Rüdiger Reitz, is an economist by training and works at the Institute for Work and Health of the German Social Accident Insurance (DGUV). His co-author, Martin Schröder, has a professional background in journalism and works at the Social Accident Insurance Institution for the Energy, Textile, Electrical, and Media products sectors (BG ETEM) in Germany.

The work of safety professionals in Germany is regulated by law, which specifies their deployment and their prior professional qualification and requires them to complete a specific training program.

The authors discuss the history of the training program, the underlying training model (which is based on a scientific study), the new competence profile for safety professionals, the competence profile for trainers, the training design, and the implementation of the course with the help of a digital learning environment. In conclusion, they report on the results of an evaluation of a recent pilot of the redesigned training program.

Aside from a new approach to teaching, which is based on modern educational principles and follows a systemic-constructivist approach, Reitz and Schröder primarily draw on the results of a longitudinal study on the effectiveness of safety professionals at companies and public institutions. With respect to the competences required for a safety professional, the study found that a systematic and methodical approach, the ability to collaborate with others, and self- and other-oriented social competences are key prerequisites for safety professionals to be able to act effectively within an organization.

Under the leadership of Reitz and Schröder, a special competence profile for safety professionals was developed. This competence profile is built on two foundations: (1) a legal provision of the German Social Accidence Insurance, DGUV Provision 2, which defines the job profile of safety professionals, and (2) the KODE competence atlas, which served as a "grid" for the redesign of the training program, enhanced by two key components: role awareness and self-reflection. The competence profile of a safety professional in Germany encompasses four central competence domains (technical, methodological, social, and personal) with a total of 13 basic competences, which were supplemented by subcompetences and descriptions of the specific form these competences take for safety professionals.

It is important, the authors point out, for the changing world of work to be reflected in the safety professional's competence profile: "Today's safety professionals embrace a comprehensive approach to occupational safety and health, they are connected to other professions and institutions, and they work to improve their company's structure and culture of prevention in a goal-oriented and systematic manner". Thus, a safety professional needs to have both profound technical expertise and strong consulting competence.

Furthermore, occupational safety and health should be an integral part of organizational structure. That is why the training program teaches participants how to build a safe and healthy structure. The training program also teaches prospective safety professionals how to identify existing patterns reflecting the organization's norms and values and the basic convictions of the people working there, and how to improve the culture of prevention in such an organization.

Based on the competence profile, the authors developed a new training design, which rests on four guidelines for teaching:

(1) design the training process around competences;
(2) enable self-organized learning;
(3) adopt a real-world approach to the training; and
(4) choose methods and media that support active, self-directed learning.

To make the program as real-world and action-oriented as possible, it was designed in a way to follow a scenario-based approach. From the beginning to the end of the program, learners progress through a series of scenarios, so-called action situations, which mirror the real tasks of a safety professional.

In terms of teaching methodology, the core element of the training program for safety professionals is a digital learning environment used at all three learning sites (classroom learning, computer-based learning, and learning at the workplace).

The authors refer to the special circumstance that a competence profile for trainers was developed almost in parallel to that for the safety professional. In this, the program pays tribute to the fact that a transformation of learning culture will only be possible if teachers professionalize accordingly. A teaching approach emphasizing self-organized and self-directed learning is bound to affect the role of the teacher as well, implying "a shift away from being 'instructors' towards becoming 'learning facilitators'". Teachers in the safety professional training program support and moderate the learning process and ensure a motivating learning environment. Flexible "learning arrangements" replace detailed instructions on how to teach a course.

The pilot phase of the redesigned training program for safety professionals in Germany was systematically evaluated in 2018. In this context, Reitz and Schröder highlight the feedback conversations with learners, which were facilitated by external moderators, as particularly helpful. These feedback sessions have become the central format for collecting learners' feedback.

The authors conclude that the redesigned training program for safety professionals is an appropriate response to the changing nature of work. It empowers safety professionals to adapt their know-how to changing situations and to play an active role in addressing this change inside their companies and institutions.

Chapter 12, concludes the book with "School Heads as Change Agents: Salutogenic Management for Better Schools". The author, Peter Paulus, adopts a different perspective: arguing from a health science perspective, his contribution explores the competences of school heads for the development of good healthy schools. Paulus is a psychologist, guest professor, and head of the Center for Applied Sciences of Health (CASH) at Leuphana University, Lüneburg, Germany. His co-author, Heinz Hundeloh, is head of the DGUV Educational Institutions Department and works at the Statutory Accident Insurance for the public sector in North Rhine-Westphalia, Germany.

Fueled by the social, economic, and technological transformation and the concomitant neo-liberal reforms to raise the efficiency of the school system, the activities and the context of school management have been seeing significant change. As key players in the design of schools, school heads have taken on new duties and responsibilities, including human resource management and development, teaching-related leadership, and organizational management and development. School heads are "change agents" for the development of schools and teaching. Success in this role requires that they are in good health.

Paulus and Hundeloh approach this issue in four steps: first, they provide figures on the health situation of students, teachers, and school heads in Germany; second, they describe health management as a responsibility of school heads and their competences; third, they explore the question of how health management may be implemented at schools; fourth, they pinpoint the link between health and educational quality as part of the concept "Good Healthy School".

Drawing on multiple recent studies, the authors begin by describing the health situation of students, teachers, and school heads.

One in ten students are involved in an accident; overweight and obesity, as well as related chronic diseases, headaches, stomach pains, back ache, and eating disorders are increasing. Bullying and psychologically caused behavioral disorders are observed more frequently. Students increasingly perceive schools as stressful places.

The teaching profession generally harbors high potential for an excessive workload and for mental and psychosomatic stress, which may lead to illnesses and even accidents. In addition, personal factors such as teachers' professional expectations regarding their self-efficacy or their knowledge of classroom management methods play a role. Taken together, a complex interaction of individual and organizational determinants leads to the physical and/or mental health situation of teachers.

The wellbeing of school heads has seen a decline in recent years. About 20% of all respondents report lower wellbeing than in 2009 and about 12% report a very low level. Yet subjective wellbeing in these studies is still higher than that reported by teachers. School heads, however, often suffer from emotional and motivational exhaustion.

The focus of Paulus's and Hundeloh's contribution is on the link between health and education: "(H)ealth, and mental health in particular, is an important resource and driver of education; that is, it is an essential prerequisite and foundation for the pedagogical work of teachers and school management in schools, but also, closely related, for the learning of school students". Their simple motto is: "Healthy school students learn better, and healthy teachers teach better".

To Paulus and his co-author, the key point is that the meaning of health at school goes far beyond the traditional understanding of health education. They understand health not as an output factor, but rather as an input factor, or a throughput factor of education.

In a second step, Paulus and Hundeloh describe the key function and responsibility of school heads when it comes to implementing health management and health promotion as part of the school's human resource and organizational development. They draw attention to the double meaning of the term health management: (1) "managing health" and (2) "managing in a healthy way". According to Paulus and Hundeloh, the number one maxim of health management at schools must be, "With health create a good school".

In a third step, the authorss describe how health management can be implemented at schools. Measures are target-group specific, including measures to develop personal resources (e.g., competence training) and measures to design schoolwork and learning environments (e.g., ergonomic school furniture but also a school culture that fosters safety and health). These measures are implemented in the areas of (1) occupational safety and health: measures to assess risks and minimize or eliminate them; (2) prevention and health promotion: school development or project management measures to strengthen and promote resources; and (3) crisis and emergency management: crisis prevention measures, crisis intervention, and post-crisis care.

In the fourth step, Paulus and Hundeloh conclude by introducing the concept of "good healthy schools" in more detail. The focus here is on the links between health and educational quality. By choosing this focus, they reject both a traditional understanding of school-based health education and an understanding according to which a school's social organization and architectural design figure as the main factors causing health or illness. Arguing from a salutogenic perspective, the authors seek to identify the extent to which designing schools according to health-related criteria (or neglecting or ignoring these criteria) affects the quality of education. From this perspective, health, embedded in the structures and processes of school work, becomes an integral part of a school that is known as a good school.

In their conclusion, Paulus and Hundeloh return to the second meaning of health management as "managing in a healthy way". It is an important task of school management to communicate health measures in a way that they can be perceived by teachers as supportive measures, as measures that help them fulfill their core responsibility of good teaching and good school education.

They recommend connecting the implementation of health at schools to a "guiding principle", such as Aaron Antonovsky's "sense of coherence". Following this guiding principle leads to an understanding of leadership with what they call "salutogenic leadership".

REFERENCES

Bartram, D. 2005. The Great Eight competencies: A criterion-centric approach to validation. *The Journal of Applied Psychology*, 90(6), 1185.

Berger, P. L. and T. Luckmann. 1973. *The Social Construction of Reality: A Treatise in the Sociology of Knowledge*. Reprint. 1st ed. 1966 USA. London: Penguin Books.

Blazsin, H. and F. Guldenmund. 2015. The social construction of safety: Comparing three realities. *Safety Science*, 71, 16–27.

Bollmann, U. 2018. Competences for a culture of prevention – Conditions for learning and change in SMEs. Chapter 6. In G. Boustras and F. Guldenmund (Eds.), *Safety Management in SMEs*, 121–141. Boca Raton, FL: CRC Press Taylor & Francis Group.

Bollmann, U. and C. Eickholt. 2019. Competences for Change. Speech at the 36th International Congress for Occupational Safety and Health (A+A), Focus Professions, 8 November 2019, Düsseldorf.

Bollmann, U., Y. Lee, Y. Seo, H. Paridon, T. Kohstall, A.-M. Hessenmöller and C. Bochmann. 2020. The development of a model and a set of leading indicators for promoting occupational safety and health. *Prevention Science*, Special issue on "Culture of Prevention", 21 (manuscript, accepted).

Cate ten, O. 2017. Competency-based medical education and its competency frameworks. In M. Mulder (Ed.), *Competence-Based Vocational and Professional Education: Bridging the Worlds of Work and Education*, 903–930. Cham, Switzerland: Springer International Publishing.

Chomsky, N. 1965. *Aspects of the Theory of Syntax*. Cambridge, MA: M.I.T. Press.

Dijk van, F., M. Bubas and P. B. Smits. 2015. Evaluation studies on education in occupational safety and health: Inspiration for developing economies. *Annals of Global Health*, 81(4), 548–560.

Dobiéy, D. 2018. Learning Trip to Corporate Creativity 4.0. Speech at IAG-Trainer Days, 6 June 2018, Dresden.

Guldenmund, F. 2015. Organizational safety culture. Chapter 19. In S. Clarke, T. M. Probst, F. W. Guldenmund and J. Passmore (Eds.), *The Wiley Blackwell Handbook of the Psychology of Occupational Safety and Workplace Health*, 437–458. Hoboken, NJ: Wiley Blackwell Publishing Ltd.

Hale, A. and K. Hovden. 1998. Management and culture: The third age of safety: A review of approaches to organizational aspects of safety, health, and environment. In A. M. Feyer and A. Williamson (Eds.), *Occupational Injury: Risk, Prevention and Intervention*, 129–165. London: Taylor & Francis.

Holland, D., W. Lachicotte Jr., D. Skinner and C. Cain. 1998. *Identity and Agency in Cultural Worlds*. London: Harvard University Press.

Mulder, M., ed. 2017. *Competence-Based Vocational and Professional Education: Bridging the Worlds of Work and Education*. Cham, Switzerland: Springer International Publishing.

Nussbaum, M. and A. Sen, ed. 1993. *The Quality of Life*. A study prepared for the World Institute for Development Economics Research (WIDER) of the United Nations University. Oxford: Clarendon press.

Nussbaum, M. 2011. *Creating Capabilities: The Human Development Approach.* Cambridge, MA: The Belknap press of Harvard University press.

Nussbaum, M. 2012. *Gerechtigkeit oder das gute Leben*, ed. H. Pauer-Studer. Frankfurt am Main: Suhrkamp.

OECD 2012. Better Skills, Better Jobs, Better Lives: A Strategic Approach to Skills Policies. Paris: OECD Publishings.

Otto, H.-U. and H. Ziegler. 2012. Erziehung und der Befähigungsansatz – Capability Approach. In U. Sandfuchs, W. Melzer, B. Dühlmeier and A. Rausch (Eds.), *Handbuch Erziehung*, 38–41. Bad Heilbrunn: Julius Klinkhardt.

Robson, L., C. Stephenson, P. Schulte, B. Amick, S. Chan, ... P. Grubb. 2010. *A Systematic Review of the Effectiveness of Training & Education for the Protection of Workers.* Cincinnati, OH: National Institute for Occupational Safety and Health. https://www.cdc.gov/niosh/docs/2010-127/pdfs/2010-127.pdf?id=10.26616/NIOSHPUB2010127.

Schein, E. 2004. *Organizational Culture and Leadership.* 3rd ed. San Francisco, CA: The Jossey-Bass Business and Management Series.

Schulte, P., T. Cunningham, L. Nickels, S. Felknor, R. Guerin, ... L. Mengler-Oge. 2017. Translation research in occupational safety and health: A proposed framework. *American Journal of Industrial Medicine*, 60(12), 1011–1022.

Sen, A. 1999. *Development as Freedom.* New York: Alfred A. Knopf, Inc.

Sprafke, N. 2016. *Kompetente Mitarbeiter und wandlungsfähige Organisationen. Zum Zusammenhang von Dynamic Capabilities, individueller Kompetenz und Empowerment.* Wiesbaden: Springer Gabler.

Swuste, P. and F. van Dijk. 2019. (Post)academic safety and health courses, how to assess quality? In P. M. Arezes, J. S. Baptista, M. P. Barroso, P. Carneiro, P. Cordeiro, C. Nélson, R. B. Melo, et al. (Eds.), *Occupational and Environmental Safety and Health: Studies in Systems, Decision and Control 202*, 785–790. Cham, Switzerland: Springer International Publishing.

Terzi, L. 2014. Capability approach: Martha Nussbaum and Amartya Sen. In D. C. Phillips (Ed.), *Encyclopedia of Educational Theory and Philosophy*, 97–99. Thousand Oaks, CA: SAGE Publications, Inc.

White, R. W. 1959. Motivation reconsidered: The concept of competence. *Psychological Review*, 66, 297–333.

Section I

Foundations

2 Reconceptualizing "Developing Competence at Work" to a Journey of Being and Becoming

Helen Bound and Kristine Yap

CONTENTS

2.1 INTRODUCTION

Developing safety and health "competence" at work is analogous to learning and developing expertise in, through, and at work. In making this seemingly simple statement, it is necessary, first of all, to unpack what we mean by "competence", "learning", and "developing expertise" particularly in the context of work. Competence is generally understood as the ability to do something well. This definition places an emphasis on performance. Competency-based systems in which much of the development of occupational safety and health is taught and accredited, work with the notion of competencies, that are often atomized, out of context, and separate technical from generic and practice from theory. In this chapter, we challenge this view, taking a socio-cultural perspective (which focuses on the interaction between the individual and their social context) placing the learners situated in the context of their work. We argue that workers need to be knowledgeable practitioners with deep know*ing* developed through engaging in work settings and exercising agency. Knowing is understood as "competent interaction with our world"; knowledge is used as a tool (Nicolini 2011, 604) as everyday action and activities are undertaken. Knowledge is not a thing to be acquired by individuals and treated as a commodity;

rather, knowing is emergent and dynamic, as we understand it. This practice-based perspective differs from the conceptualization of knowledge in competency-based training, where knowledge is often conceptualized as static, unquestionable, and as separate from enactment. To understand the journey of workers becoming knowledgeable practitioners and their exercise of agency, it is necessary to understand the context at specific work sites, in relation to industry, occupational practices, national law, and legislative requirements.

The potential for change is at the heart of developing cultures of prevention in relation to safety and health. In considering competence and developing cultures of prevention, the question becomes, in what ways can competence contribute to cultures of prevention? In this chapter, we draw on the theoretical perspectives of being and becoming to consider competence in relation to developing cultures of prevention. This requires a repositioning of notions of competence as enacted in competency-based training (CBT) systems. Furthermore, we suggest that being and becoming a practitioner in relation to safety and health is intrinsic to, not separate from, the vocational/professional practices of the practitioner and the contexts in which practitioners work and live. There is a constant process of negotiation between individual and the social context, inclusive of developing "cultures of prevention". Being and becoming invokes the metaphor of journey; a journey that never ends, a journey in which identities are always evolving. That work identities are constantly negotiated and renegotiated and are part of workplace learning is well supported in the workplace learning literature (see, for example, Billett et al. 2006; Billett et al. 2008; Eteläpelto 2015). The renegotiation of work identities is necessary for changes in work practices and organizational change (Eteläpelto 2015). Being, becoming, the negotiation and renegotiation of work identities are active social processes, requiring the exercise of agency to influence changes in work practices, often necessary where a shift to a culture of prevention is being established.

Identity and agency are intrinsically linked. The authors argue that individuals are not merely pawns within structures, power relations, and cultural norms, discourses and policies; within the affordances of their circumstances, individuals have the ability to take actions, make decisions, and participate in activities. This is the exercising of agency. The nature of actions, activities, and involvement is dependent on individual biographies and circumstances that inform identities. Identities constantly evolve over the journey of a lifetime and are mediated by the social worlds in which individuals have been part of, both currently and historically.

Identity, agency, being, becoming seem a world away from traditional notions of competence as static attributes of knowledge, skills, and attitude. Not surprising, as the concepts are based on quite different views of knowledge and of the world. The former comes from a socio-cultural perspective and the latter is a traditional notion of competence that comes from cognitive and behaviorist understandings. These different approaches are important to understand as they name the world and how it is understood and that in turn informs our thinking and actions (Freire 1977). In thinking about being and becoming (Holland et al. 1998), it is necessary to also understand the contexts in which identity and agency are exercised. Whereas a traditional competency approach considers the individual as isolated from their context.

Given that safety and health is about life and death, about wellbeing or the lack of it, in this chapter the authors argue that holistic understandings of individuals in their contexts and their life histories is necessary in order to develop competence that contributes to a culture of prevention. For example, in a workplace involving heavy machinery in Singapore, there are workers from many nationalities and cultures, where Singaporeans work closely with other migrant workers from neighboring countries: India, mainland China, and Malaysia. The local on-site supervisors would often be frustrated with foreign team members not wearing steel capped boots, helmets, and other personal protective equipment and with those who operated the forklifts zapping across the workspaces at high speeds. To address this "problem", a member of the human resource team focused on addressing the performance gaps in the competencies foreign equipment operators appeared to be lacking in. However, the human resources (HR) personnel (we shall call him David) was at the time undertaking a master's course with a unit in workplace learning. On understanding more about the processes and perspectives involved in supporting and designing for workplace learning, David realized that the deficit approach of assuming that the foreign operators lacked skills needed to be challenged. He commented in an assignment (data from an Institute for Adult Learning and CRADLE, Nanyang Technological University project on Dialogical Teaching by Bound et al. 2019) that:

> I was … influenced by preconceived stereotypical ideas of the foreign operators' work practice. These had blindsided me from their strengths and competences, which were equally important, as these are personal factors that would also influence and shape their learning interventions. I realised that the stereotypical perception that they were novices had to be challenged, as their behaviors were not due to their lack of skills, but due to the culture and the work environment they had back home. This subsequently led me to reframe my perspective and improve the roadmap [the learning intervention to establish a culture of safe practices] by looking at strategies that capitalize on the strengths of the learners and supervisors who are their mentors.
>
> **(David (a pseudonym): permission to use this material was sought and gained from "David". See Bound et al. 2019 for details of this project)**

David also realized that there was a lack of understanding of each nationality's working culture, and a lack of trust in the competence of the group of foreign operators by the local supervisors and operators even though the operators might have years of experience back in their home country. He noted that it did not help that there was limited face-to-face interaction between the different groups, and limited time for coaching and reflection sessions with mentors and supervisors.

This story is illustrative, not only of deep changes in the perspectives and understandings of David, but of the importance of paying attention to context and life histories. The foreign operators brought with them their own practices and experiences. In their countries of origin, safety standards are often practiced differently with limited legislation in relation to safety. For example, their ability to drive equipment at high speed weaving paths outside the designated lines demonstrates considerable competence in handling the vehicle, but it is deemed as a safety hazard in Singapore, where there are established safety regulations and practices. So the driving competences of the operators were not a lack of driving competences, quite the opposite, but came from a different understanding of the practices and the job. Now the "problem" is defined and understood differently, not as one of a deficit of skills, but as far more complex. What needed addressing was an appreciation and understanding by each group of different practices. Such an understanding then changes the nature of the intervention and explanation when working with practices that are not conforming to regulations and site practices. The lack of trust between groups is a workplace culture issue and a cultural issue. Such an understanding of the context, along with some understanding of life histories of the individuals led to a holistic intervention to address the issues. An approach of addressing "gaps" in competencies would not have addressed the fundamental issue.

Changes in identity and thus of being and becoming are evident in this story. The first is that of David who was actively re-negotiating his perspective and understandings, that is, his meaning-making of the situation. Ethnographic data from a study on dialogical inquiry as a pedagogical practice (Bound et al. 2019) unveils this shift in David's stance and in the execution of his role as an HR manager. He exercised considerable agency in changing the typical HR response and practices from a deficit approach based on traditional understandings of competencies to a holistic approach, delving deeper to understand the problem, and addressing contextual and individual needs. Supervisors and operators across the different nationalities also renegotiated their identities and work practices as a result of the intervention to improve safety, such as sharing of stories, putting individuals in the shoes of others, and so on, to develop cultural understandings and a deep understanding of safe work practices.

We begin by unpacking the notion of "cultures of prevention", then move onto a discussion of competence and its relation to being and becoming. A discussion of agency and competence follows before we provide specific examples of integrating classroom and work for learning and developing cultures of prevention.

2.2 DEVELOPING CULTURES OF PREVENTION

How to develop cultures of prevention? In considering this question, it is helpful to unpack what we mean by "cultures". Bishop et al. (2006) point out that while the term "culture" as it relates to organizations is contested, as is the extent to which cultures can be managed, there is some general consensus that it is inclusive of implicit values, assumptions, and norms that are generally invisible; culture is also inclusive of what is visible, namely, practices and artifacts (Bishop et al. 2006). Culture mediates relationships, creating the context for social interaction and provides affordances for applying and creating knowledge (DeLong and Fahey 2000). Brown and Duguid (1991) point out that a strong culture reduces uncertainty and provides means for coordination and control.

Changing values and identifying assumptions is difficult, but De Long and Fahey (2000) suggest that changing practices can lead to change. For example:

> (T)he ways in which departmental meetings are conducted strongly influence the likelihood of a group's generating new knowledge or leveraging its existing knowledge.

Are differences of opinion encouraged and respected, or routinely discounted by group leaders? Is conflict managed constructively or is it suppressed or smoothed over?

(De Long and Fahey 2000, 115–116)

The routine discounting of ideas reflects a culture of fear of loss of command and control by management. Such a culture cannot be easily replaced with a culture that encourages and respects differences of opinion as the beliefs and values behind these different practices are poles apart. As Bishop et al. (2006) point out, altering cultural *practices* can only ever be a part of the culture change process as there needs to be alignment between changes in practices *and* beliefs, norms, and assumptions.

A belief that humans lack knowledge and must be topped up with the requisite competencies (a deficit approach) is likely to give limited discretion to employees, whereas a belief in human capacity for sense-making and co-creation of knowledge likely contributes to opportunities for discretionary effort. It would seem then, that developing cultures of prevention may be difficult. For example, Billett (2011) recounts how work practices in mine sites are shaped by the masculine culture of the workplace and the mining community; this masculine culture was often manifested in unsafe work practices. Only when there are significant events such as deaths, major injuries, or severe ill-health, do miners engage actively in a questioning of the hegemonic masculine culture and even then, there is frustration at their "inability to influence other miners' behaviors and practices" (l.c., 65). Given these considerable difficulties in changing cultures, how else might we think about developing cultures of prevention?

When considering culture, structure influences culture and vice versa. Structure is evident in more than reporting lines; it is evident in the flow of work that structures work activity; it is evident in standard operating procedures, which provide a process and structure for carrying out routine tasks; it is evident in how meetings are conducted and if multiple voices are heard, or silenced, and so on. In these ways, beliefs and assumptions are embedded in various structures in an organization. There is considerable discussion in the literature about structure and agency (see Eteläpelto 2015). The authors of this chapter argue that individuals are not powerless within structures and cultures that work against safe practices. While agency may be bounded (Evans 2002), there is generally some space for the exercising of agency and thus contributing to the potential for dynamic changes in practices. This interaction between the individual and the "social world" (e.g., specific work settings) is made clear by Billett:

The social world requires individuals' agency to actively remake and transform its practices, while individuals need the social world to provide access to knowledge that is sourced in history, cultural practices (i.e. occupations) and manifested in particular instances of the practice of that work.

(Billett 2011, 61)

From a socio-cultural perspective, knowledge, beliefs, and values are embedded in tools (e.g., nail gun, smartphones, ways of thinking where everyone is expected to contribute at meetings and different perspectives are appreciated) and the history of development of those tools, how they are used, and when and why they are used. Cultural practices associated with ways of working and thinking, beliefs, assumptions – are all embedded in tools used every day. In reality, different individuals use tools in different ways and thus are constantly either remaking cultural practices as their thinking and use of the tool(s) reproduces practices or they contribute to what are often small changes in practice as they apply the tools in – what may be quite subtly – different ways. For example, Yap (2017) argues that sharing and communication can positively guide and shape the enactment of safe and healthy work practices, enabling reinforcement and also the creation of new knowledge. In the petrochemical plant in Singapore where Yap (2017) undertook her ethnographic study of learning safety and health practices in the workplace, deliberate institutional interventions are necessary to set the stage to encourage and develop collective learning. An example from an interviewee, who we will call Aamil, explains the change from limited participation and engagement in work group meetings to full engagement and participation – testimony to the potential of genuine intent to change practices. As English is not Aamil's first language square brackets inside the quote explain what he is saying.

> In those days, two people like to go up [to the front of the meeting] but now the mentality of workers are different, everybody rushing to go up and share their experience. This is where we are seeing a culture, a good culture being practiced down to the workers [by everyone]. So that is where we share all these incidents or procedures, so it is very good so called from the workers so one thing over here, we [are] working like a family [i.e., workers support each other].

(Aamil, interview quoted in Yap 2017, 88)

Such active engagement in practices (Schatzki 2012) is necessary to access knowledge that is embedded in work practices, which eventually develops into a distinct organizational culture and identity for safety. As individuals engage in goal-directed activities, there is a cognitive legacy, a change in cognitive structures or knowledge. "It will be individuals' agency that will adapt that knowledge in new ways and to novel circumstances in workplaces, thereby remaking and transforming those practices, as they learn themselves" (Billett 2011, 64).

Aamil's words, along with other evidence he provided that there is no space to share here, indicate that agency does not just belong to the individual, it is also collective. As Aamil states, the plant works "like a family", there is a sense of belonging, an identity with the work and other workers, which is important in encouraging the preparedness to share among the work community. The values and beliefs underlying such sharing, indicate that management genuinely believes in the importance of such sharing, and believes that the voice of each individual is important. Adopting the practice of sharing experiences and issues contributes to changes in identity, what it means to *be* a worker in this particular plant – there is a sense of looking out for each other that is behind the sharing. At the same plant, a worker, Rauf, comments, "Share, yes, they will feel like we, togetherness, grow together" (Yap 2017, 118). In this plant of multiple nationalities there is a deliberate putting together of different nationalities to learn from and support each other to not only work safely, but also to improve performance. This is important when many of these workers are far from home and support structures. Such active engagement and interaction establish a collective competence where there is a shared understanding of the work (Sandberg 2000). Participants are "being involved in the ongoing meaning-making" (l.c., 59), giving them a specific direction and orientation in performing the work, "shared understanding is cultivated, refined and maintained through their ongoing involvement in collectively making sense of their work" (ibid.).

The argument so far has been that in developing cultures of prevention, we need to pay attention to practices, beliefs, and values. That active engagement in practices is what contributes to active meaning-making, a sense of identity and agency. Thus, there is a holistic understanding of learning and development that embodies identity and agency, a sense of being and becoming, what it means to *be* a particular vocation, professional, a worker in a particular setting. Such engagement establishes collective competence – critical for developing a culture of prevention. Developing competence can be about being and becoming, about engagement and increasing capacity to act differently in various and varied work settings. How does this understanding of being and becoming sit with different concepts of competence and its potential for contributing to developing cultures of prevention?

2.3 DEVELOPING COMPETENCE AS A WAY OF BEING AND BECOMING

In referring to the competence of professionals, competence is generally understood as the ability to do something well. However, with the advent of CBT systems, there is potential for confusion and overlap between the common understanding of competence and the way it is conceived of in CBT. Learning theories underpin different conceptualizations and we need to surface these to better understand the changing perspectives and meanings of competence. It is also important to understand these different theoretical lenses as they inform approaches and practices to developing and supporting cultures of safe working practices (Bound and Lin 2013). For example, Sandberg (2000) refers to rationalistic and interpretive approaches, Mulder et al. (2007) to the behaviorist, generic, and cognitive approaches, and Delamare, Le Deist and Winterton (2005) write about behavioral, functional, multi-dimensional, and holistic approaches, the latter to develop a proposed typology of competence.

Any practitioner involved in supporting learning for cultures of occupational safety and health (OSH) practices will likely draw from a range of theoretical perspectives. But as illustrated in David's story (where he moved from a predominantly behaviorist, human capital perspective to a socio-cultural perspective that embodied being and becoming) most of us hold a dominant perspective, whether consciously or unconsciously. As stated in the introduction, it is important to understand our stance, our perspective as this mediates our actions. In the following paragraphs, we outline competence from the perspective of the major learning theories and then discuss alternative understandings of competence in relation to being and becoming.

Behaviorism makes the claim that the world exists outside the individual, it distinguishes between the knower and his/her world. Behaviorists seek evidence by observing behavior. Consequently, learning is understood as occurring by defining specific observable, predefined behaviors (Schuh and Barab 2008). It is observable job performance that is the focus in this perspective. The individual and the world are considered as separate entities. Behaviorism claims an objectivist perspective based on logic and deduction assuming the existence of an objective reality, independent of the human mind (Sandberg 2000). Competence is described from this perspective as "consisting of two independent entities – prerequisite worker attributes and work activities" (l.c., 11). Therefore, the term "competency" is used rather than competence. The former is a narrow understanding of competence that assumes specific predefined attributes. The acquisition metaphor of learning (Sfard 1998) aligns with this perspective.

Older cognitive approaches also typically set up a dualism. Individual mind(s) and the world are considered to be separate entities; that is, that individual mental processes are used to develop understanding, but context is external to the individual. From a cognitive perspective, knowledge is learned and changes in knowledge make changes in behavior possible (Woolfolk and Margetts 2007). While older views of cognition emphasize the acquisition of knowledge, newer approaches stress the construction of knowledge as individuals pay attention to what is happening around them. In CBT documentation there is a separation of "underpinning knowledge" and skills or performance, i.e., the doing. This suggests that the acquisition approach is being applied. Mulder et al. (2007) suggest that cognitive competences include specialized cognitive competences, such as social and emotional competences. These social competencies are more commonly known as generic competences.

Interpretative approaches move away from the dualism present in the approaches outlined above – believing the world and mind are socially constructed. For Sandberg (2000, 11), "person and world are inextricably related through persons' lived experience of the world". This has significant implications for the way in which competence is understood, not as separate entities of worker and work, but as "one entity through the lived experience of work. Competence is seen as constituted by the meaning the work takes on for the worker in his or her experience of it" (ibid.). From the interpretative perspective, competence is about workers (individually and collectively) developing understandings of the work, as opposed to an attribute-based-only perspective that assumes that the mind and the world are separate entities. This perspective allows for a developmental perspective of individuals and collectives as they interact with their world. However, the ways in which competence is actually used is dependent on the ways in which work is organized, the degree of autonomy workers have, the extent of participation, the characteristics of the task, and the nature and degree of feedback (Ellström 1997).

Consistent with our argument in the previous section, we claim that competence is not just acquiring knowledge or updating skills (Bound and Rushbrook 2015). We argue that competence in the sense of doing something well is developed individually and collectively. Development of competence is through engagement in work practices, which in turn are mediated through the organization and design of work, power relations and decision-making processes, and who has access to knowledge. In addition, competence development involves identity work. Identity work refers to the evolving journey of identity development and that informs what individuals care about, shaping agency that is often dependent on how individuals position themselves and how others and the social relations of the context position them

(Holland et al. 1998). This suggests that current CBT approaches limit possibilities for contributing to cultures of prevention, as they are based on dualisms that separate mind from body, theory from practice, and the individual from the collective. More specifically, CBT places a focus on the *acquisition* of skills and knowledge, held by the individual, rather than the social construction and creation of knowledge, and of contributing to evolving practices and thus cultures of prevention. It seems more useful to understand competence as doing something well, and as requiring agentic engagement in practices; that as "individuals practice agency, they construct knowledge, using metacognitive and reflective processes in managing learning and problem-solving" (Eteläpelto 2015, 53). In exercising and practicing agency, we can see that being and becoming for the individual and the collective requires learning to learn meta-cognitive capabilities (Bound et al. 2018a). This aspect is important in developing cultures of prevention as it is future-oriented (Bound et al. 2016) and builds in capacity for continual learning.

2.4   DEVELOPING COMPETENCE AND AGENCY

Yap's case study of "learning safety" in a petrochemical plant in Singapore (2017) illustrates how learning is much more than acquiring technical knowledge and skills. This study suggests that learning to do something well and, at the same time contribute to a culture of prevention, involves individuals engaging in:

- exercising their agency (confident in taking initiative);
- dialog, sharing, exchange, and knowledge creation;
- meta-cognitive, reflective processes;
- embodied learning;
- exercising discretion:
  - making judgments and self-reasoning rather than acquiescently following guidance from others;
  - actively utilizing knowledge;
- responding effectively to circumstantial requirements; and
- collective practices.

All of which are processes involved in being and becoming an OSH practitioner, contributing to a culture of prevention.

Involvement in collective practices, in dialog, sharing, and exchange, and being agentic in these situations and in others are expounded on above. In the following, we provide examples of using discretion, embodied learning, and responding effectively to circumstantial requirements.

Exercising agency to use critical assessment in everyday work despite prescribed rules and standards are important in responding to circumstantial requirements, as described by Peng in the following quote:

> SOP is only a Standard Operating Procedure but those danger[s] that you do on the way [unexpected situations, complex, non-standard problems], it won't be inside the SOP. That's [SOP] only the process, emergency, all the operating parameters but when

you are in the field, like maybe space limitation, the height, all these, you won't be so detailed that every small, small thing is all inside.

(Peng, interview quoted in Yap 2017, 97)

Peng explains that SOPs only cover standard, non-complex situations and processes, so that when he experiences a more complex problem, the SOP is not enough; there is much more "inside" the problem than an SOP can hope to cover. Working safely in this perilous petrochemical work setting requires exercising judgment and discretion outside of SOPs.

Loh, who works in a different part of the petrochemical plant cites a typical situation in his everyday work in on the instrumentation panel, illustrative of the engagement of all senses, the intuitive self, and the observation of the physical and contextual requirements to guide and make his own judgments and decisions. Speaking about his job role as a control room operator:

but sometimes you will need to work on your own discretion. Cannot be every time follow what other people say, because the parameters change. Suddenly this equipment, suddenly, you thought it's actually nothing inside, it's quite safe. But then when you open up hot oil come out. Then you get panic and all this thing. So, you also must listen: what am I doing? I need to know a lot of things inside [needs to know what is happening inside the pipe].

(Loh, interview quoted in Yap 2017, 97)

Meta-cognitive processes are evident in Loh's observation. Ensuring safe practices means trusting your own senses; consciously using the ability to hear what is happening inside the line and drawing on knowledge about what is going on in relation to that line, and no doubt more theoretical knowledge about flows all contribute to keeping Loh, and those around him, safe.

Engaging in collective practices is a process of approaching, for example, problem solving collectively, and at times creating collective practices along the way, as indicated in Ming's account of how his team and the contractors worked together to repair a pump:

Because the contractors after, for example, we isolate one pump or isolate the strainer and maybe the changer, the equipment, our equipment and then the isolation part maybe have some problem... and then the contractors cannot carry on their work. Because our part is not done good enough, their part will not be safe to proceed with their work and then we got to do our part. Once we hand over, sometimes we overlook some area, cannot see one, and then they will highlight to us. And then, if still not safe, discuss with contractors. The contractors may have some better good idea. And then sometimes maybe if for example cannot remove oxygen, cannot remove fuel, maybe the fuel is present, maybe we can remove the heat, we can do a cold cut.

(Ming, interview quoted in Yap 2017, 93)

Ming explains that if one group of contractors does not undertake their work according to the required safety standards, it will not be safe for the next group of specialized contractors to do their work. These situations require close communication between different groups of contractors. His example of doing a cold cut means even if there is still a little residual fuel in the line, it will be safe as there is no heat to ignite the fuel. Seemingly simple things such as understanding whole processes, rather than just parts, requires and also develops collective understanding. How workplace activities are structured is also pertinent to facilitate the exchange of knowledge that contributes to the culture of prevention which must be part of management practices. Peng gives an example:

> Sometimes certain things inside [inside the pipe or equipment that cannot be seen] we don't know what's, how it operates until the maintenance people open up then they show to us how is inside like during our turnaround, shutdown maintenance then we get to see what's inside the tower.

(Peng, interview quoted in Yap 2017, 92)

This example of boundary crossing where different teams and areas of expertise are working together, sharing knowledge, and creating new knowledge for individuals and the collective illustrates the importance of understanding whole processes. Understanding the whole process is necessary in ensuring workers are cognizant of the work they do and its effect on others down the line or those who are in the same space after them. Particularly in workplaces with a diverse work community, providing communicative spaces is critical, not only for clarification and validation of what is learned and what is being interpreted, but instrumental to understanding how those personal and agentic influences mediate individual and group thinking and doing.

Agency, work, and learning are intricately entwined with and embedded in each other. This has implications for a focus on developing cultures of prevention, unstructured learning in workplaces, and affordances for where and how accredited, structured learning takes place.

2.5   INTEGRATING CLASSROOM AND WORKPLACE LEARNING

Assumptions that what is learned in the classroom is seamlessly "transferred" to the work have long been shown to be flawed (see, for example, Tennant 1999; Evans et al. 2011); yet, despite this, there continues to be a reliance on classroom delivery that is treated as quite separate from the "real" world of work. Be it pre-service qualifications or part of continuous professional development, improved integration of OSH practices into the design of learning, using authentic problems, challenges, and tasks to close the gap between classroom or online learning environments and work needs attention. Bound, Chia, and Lee (2018) argue that different spaces afford opportunities for different pedagogical tools. Different settings such as classrooms, simulators, skills practice, and so on are distinguished and shaped by different purposes and outcomes of learning, which then lend themselves to different pedagogical tools that vary in types and degrees of complexity. Another important aspect facilitating application of OSH practices in everyday work is the development of deep understanding to enable application (Tennant 1999). Through multiple authentic experiences (which may, for example, take place in classrooms but might be drawn from the workplace, and/or take place in work settings, and/or might be challenges arising from the literature that relate to real issues), access to and engagement with information and co-construction of knowledge help learners to develop deep understanding and appreciate the importance of working safely. Deep understanding is beyond learning the steps to follow, the "how to"; deep understanding is about being able to apply the principles behind the how to. Such understanding enables the integration of theory and practice, and problem solving in and across different contexts (Bound et al. 2019).

A key focus of Yap's (2017) study is the importance of the workplace as a space for developing safe practices and cultures of prevention. For example, as a supervisor, James observes that "Hands on actually patrol yeah because we see the actual one, actual situation. Sometimes we cannot imagine. This thing people do this way, I never imagine" (in Yap 2017, 98). James highlights the importance of being physically involved (actually patrolling and seeing the real situation) and having the opportunity to directly observe as he learns to hone his skills in everyday practice, rather than relying on classroom lectures.

Workplaces provide rich learning experiences, where individual perceptions are continuously negotiated and shaped by competing discourses within the physical and social contexts of workplaces. The appropriateness and validity of knowledge learned as well as the quality and legitimacy of guidance is paramount for ensuring collective understanding of OSH practices.

In designing "training" for safety, we need to ask what it is we want our practitioners to be and become, rather than using the acquisition metaphor of learning (Sfard 1998) where learners "acquire", and are given static knowledge that they hold but do not necessarily know what to do with. What we want our practitioners to be and become involves developing graduates who think safety, are safety conscious, as it is part of their being, who know and do safe practice automatically, and yet are agentic (take action) to ensure safe practices. In other words, it shifts from competency as knowledge, skills, and attitudes to competence and learning as the "transformation of understanding, identity and agency" (Edwards et al. 2002, 532).

Although important, training is but one aspect of developing cultures of prevention. Important work has to be undertaken at the worksite. Cultures of prevention are not about separate individuals, but are collective, involving a distributed know*ing* across individuals, the socio-material conditions, and the leadership in the organization.

2.6   CONCLUSION

The relation between competence and learning is generally understood as a set of competences, either at a workplace or national level, and workers are expected to learn these or to already have "acquired" them. Changing and developing cultures, be it around OSH practices or any other content, involves a change process that is far more complex than having sets of competences understood as knowledge, skills, and attitudes, as illustrated in David's story above. Change processes may be driven entirely from the top down (i.e., from management to workers), from bottom-up (by workers, supported, for example, by their union) or by a combination of both approaches. Again, it is important for practitioners driving change to name and make visible their assumptions about change processes; do they, for example, take a more democratic approach like the one advocated in this chapter or a more authoritarian approach, enforcing compliance?

Change processes inevitably involve learning, so we might ask, is the change process being conceived as a developmental, transformative process or is a functional, competence approach being used?

Developmental, transformative approaches pay attention to being and becoming, to identity and agency, valuing the contribution of each individual and the collective. Yap's study indicates that the notion of working as a family, the need to be empathetic and watch over one another's safety, and wellbeing at work are important for OSH practices (2017). Developmental approaches where peers and management act as guides and discussants influence how workers connect their professional identities with others, including those they receive guidance from (i.e., expert others). Agency thus becomes necessary in understanding and appreciating risks that will encourage deliberate efforts to enact in OSH practices. Worker learners engage in a process of meaning-making and make connections with both their professional and personal identities. How they establish congruency and seek mutual agreement across different cultural and personal attributes as individuals, while they engage socially with other co-workers can shape the outcome of OSH practices. OSH practices need to be understood in terms of the distinct circumstances of the workplace in which workers continuously engage in a process of seeking situational understanding, making intuitive decisions and judgments (Beckett and Hager 2002; Cedefop 2014) that are necessary for the enactment of safe and healthy working. Seemingly, as individuals act, they also set the stage to construct particular structures and events which eventually guide safety practices and performance (Weick 2009). It is also through this meaning-making process that they deepen their knowledge and practice contributing to their journey of being and becoming. This process of "sense-making" that individuals engage in means that they continuously question their capacity to act. As they seek to improve current expectations based on new experience gained through learning and reflection, safety culture is cultivated to improve system outcomes (Weick 2009).

Essentially, learning to work safely and healthily entails more than just developing technical and theoretical knowledge. While rule following still prevails as a conventional approach to working safely, there are also circumstances which require individuals to act beyond those prescribed rules or established requirements. Learning as experienced by different workers suggests it is determined by how individuals interact with and respond to those institutional and social imperatives in the workplaces, as well as their personal histories and epistemologies.

Developing "cultures of prevention" requires the exercise of personal agency to transform existing practices. Traditional CBT separates technical and generic skills, theory, and practice from each other and focuses on individuals *acquiring* skills and knowledge where everyone is expected to receive the same experience of learning these skills and knowledge. Yet, individual biographies and circumstances mean that each person engages and makes sense differently – "even the most uniform social experience will be subject to interpretation, construal and construction" (Billett 2011, 63). Once individuals have attained these skills and knowledge, it is a considerable leap for them to actively engage in change processes. If we understand that the relationship between individuals and the social is relational, not separate, then we are in a powerful place in which to consider the development of "cultures of prevention".

REFERENCES

Beckett, D., and P. Hager. 2002. *Life, Work and Learning: Practice in Postmodernity.* London: Routledge.

Billett, S. 2011. Subjectivity, self and personal agency in learning through and for work. In *The SAGE Handbook of Workplace Learning*, eds. M. Mallcoh, K. Cairns, K. Evans, and B. N. O'Connor, 60–72. Los Angeles, CA: SAGE Publications.

Billett, S., T. Fernwick, and M. Somerville. eds. 2006. *Work, Subjectivity and Learning: Understanding Learning through Working Life.* Dordrecht, The Netherlands: Springer.

Billett, S., C. Harteis, and A. Eteläpelto. eds. 2008. *Emerging Perspectives of Workplace Learning.* Rotterdam: Sense Publishers.

Bishop, D., A. Felstead, A. Fuller, N. Jewson, T. Lee, and U. Lorna. 2006. *Connecting Culture and Learning in Organisations: A Review of Current Themes*, Learning as work research paper, No. 5 (June).

Bound, H., and M. Lin. 2013. Developing competence at work. *Vocations and Learning* 6(3), 403–420.

Bound, H., and P. Rushbrook. eds. 2015. *Towards a New Understanding of Workplace Learning: The Context of Singapore*, 36–53. Singapore: Institute for Adult Learning.

Bound, H., A. Chia, and A. Karmel. 2016. *Assessment for the Changing Nature of Work: Cross-Case Analysis.* Singapore: Institute for Adult Learning.

Bound, H., A. Chia, and W. C. Lee. 2018. Spaces and spaces 'in between' – Relations through pedagogical tools and learning. In *Integration of Vocational Education and Training Experiences Purposes, Practices and Principle*, eds. S. Choy, G.-B. Wärvick, and V. Lindberg, 243–258. Singapore: Springer.

Bound, H., S. Sadik, K. Evans, and A. Karmel. 2019. *How Non-Permanent Workers Learn and Develop: Challenges and Opportunities.* London: Routledge.

Bound, H., T. S. Chee, A. Chow, W. Xinghua, and C. K. Hui. 2019. *Dialogical Teaching Investigating Awareness of Inquiry and Knowledge Co-Construction among Adult Learners Engaged in Dialogic Inquiry and Knowledge (Co)Construction.* Singapore: Institute for Adult Learning.

Brown, J. S., and P. Duguid. 1991. Organizational learning and communities-of-practice: Toward a unified view of working, learning and innovation. *Organization Science* 2(1), 40–57.

Cedefop – European Centre for the Development of Vocational Training. 2014. *Terminology of Vocational Training Policy: A Multilingual Glossary for an Enlarged Europe.* 2nd ed. Luxembourg: Office for Official Publications of the European Communities. https://www.cedefop.europa.eu/files/4117_en.pdf (accessed December 12, 2019).

Delamare Le Deist, F., and J. Winterton. 2005. What is competence? *Human Resource Development International* 8(1), 27–46.

De Long, D., and L. Fahey. 2000. Diagnosing cultural barriers to knowledge management. *Academy of Management* 14(4), 113–127.

Edwards, R., S. Ranson, and M. Strain. 2002. Reflexivity: Towards a theory of lifelong learning. *International Journal of Lifelong Education* 21(6), 525–536.

Ellström, P. E. 1997. The many meanings of occupational competence and qualification. *Journal of European Industrial Training* 21(6/7), 266–273.

Eteläpelto, A. 2015. The role of identity and agency in workplace learning. In *Towards a New Understanding of Workplace Learning: The Context of Singapore*, eds. H. Bound, and P. Rushbrook, 36–53. Singapore: Institute for Adult Learning.

Evans, K. 2002. Taking control of their lives? Agency in young adult transitions in England and the New Germany. *Journal of Youth Studies* 5(3), 245–269.

Evans, K., D. Guile, and J. Harris. 2011. Rethinking work-based learning: For education professionals and professionals who educate. In *The SAGE Handbook of Workplace Learning*, eds. M. Mallcoh, L. Cairns, K. Evans, and B. O'Connor, 149–161. Los Angeles, CA: SAGE Publications.

Freire, P. 1977. *Pedagogy of the Oppressed.* Harmondsworth; New York etc.: Penguin Education.

Holland, D., W. Lachicotte Jr., D. Skinner, and C. Cain. 1998. *Identity and Agency in Cultural Worlds.* London: Harvard University Press.

Mulder, M., T. Weigel, and K. Collins. 2007. The concept of competence in the development of vocational education and training in selected EU member states: A critical analysis. *Journal of Vocational Education and Training* 59(1), 67–88.

Nicolini, D. 2011. Practice as the site of knowing: Insights from the field of telemedicine. *Organization Science* 22(3), 602–620.

Sandberg, J. 2000. Understanding human competency at work: An interpretative approach. *Academy of Management Journal* 43(1), 9–25.

Schatzki, T. R. 2012. A primer on practices: Theory and research. In *Practice-Based Education: Perspectives and Strategies*, eds. J. Higgs, R. Barnett, S. Billett, M. Hutchings, and F. Trede, 13–26. Rotterdam: Sense Publishers.

Schuh, L. K., and S. A. Barab. 2008. Philosophical perspectives. In *Handbook of Research on Educational Communications and Technology*, eds. J. M. Spector, M. D. Merrill, J. van Merrienboer, and M. P. Driscoll, 67–82. New York: Lawrence Erlbaum Associates.

Sfard, A. 1998. On two metaphors for learning and the dangers of choosing just one. *Educational Researcher* 27(2), 4–13.

Tennant, M. 1999. Is learning transferable? In *Understanding Learning at Work*, eds. D. Boud, and J. Garrick, 165–179. London: Routledge.

Weick, K. E. 2009. *Making Sense of the Organization.* Chichester, PA: Wiley.

Woolfolk, A., and K. Margetts. 2007. *Educational Psychology.* Frenchs Forest, NSW: Pearson Education Australia.

Yap, K. 2017. *Learning Safety in the Workplace: A Case Study of Petrochemical Workers in Singapore.* Doctor of Education thesis. School of Education and Professional Studies, Griffith University, Australia.

3 Culture of Prevention and Digital Change
Five Theses on Work Design

Just Mields and Ulrich Birner

CONTENTS

It is autumn and we are on an express train from Munich to Cologne. Everyone in the open coach is holding a smartphone or tablet, and some people are editing documents on laptops. Their faces are illuminated by the screens. No one notices that night is beginning to fall in the fields rushing by. All their attention is focused on the digital reality. On a world that multiplies our analog world. Everything and everyone are everywhere at all times. Social media, games, and management tools create contact with thousands of "friends", enabling fantasy worlds to be conquered and corporations to be guided. No place is so remote that we cannot contact a customer or a colleague or resolve some problem.

The first smartphone – an iPhone – was launched on 29 June 2007. Ten years on, almost everyone carries one of these small computers with them. Our living conditions are changing radically due to digitalization. Is it too late to assess the impact on people, identify risks to our safety and health, and take action in time?

3.1   INTRODUCTION

Here, we define digital transformation as a disruptively rapid, dynamic process of change, driven by digital technologies and affecting all levels of society including businesses. The impacts are the subject of lively discussions in countless publications and various expert committees. However, the academic standards underpinning the debate are poorly developed, so meaningful conclusions are difficult to pin down. The contributions to the debate are frequently vague about the terms of reference and concepts of *digitalization, automation, smart factory, globalization, Internet of things, demographic change*, and so on. Often, they are informed by extrapolations of present-day digital developments to an indeterminate time in the future, or they combine the opinions and assessments of experts and stakeholders from different disciplines or with different roles.

The essence of all these contributions is that the digital transformation influences all elements of the work system (BMAS 2015, 2017; Diebig et al. 2018): artificial intelligence is bringing about new qualities in human-machine interaction, a shift towards mental work requirements, and an intensification of work. In the affected work tasks or job types, higher standards are required in terms of individual competence and the quality of cooperation. Activities are becoming more transparent, easier to monitor, and easier to supervise. Working time and location are becoming more flexible, hand in hand with the breakdown of the boundaries between private and working life. Digitalization is encountering a social change in which life plans are becoming more individual and more varied, classic role models are becoming less rigid, and values and expectations of work are being transformed.

This also presents new challenges in the development of a culture of prevention in occupational safety and health. Bollmann (2018, 125) attests to the importance of the necessary individual competence and the increasing attention on the activation and development of competences in social systems. From a systems-theoretical perspective (Simon 2009), however, the complementary view of organizational competences appears necessary. Individual competences can contribute to organizational action only to the extent provided and allowed by the culture of the organization. Organizational competences are defined as the self-regulating capacity of an organization to continuously and successfully adapt to the changing challenges of the world around it. This primarily means patterns of collective problem-solving that are not only informed by the particular corporate culture but can also contribute to its development. With regard to technological change, Schein observed the following as far back as 1985: "If the technology changes in a substantial fashion, the organization or occupation not only must learn new practices but must redefine itself in more substantial ways that involve deep cultural assumptions" (Schein 1985, 26).

Digital transformation therefore requires attention to be paid not just to the technical changes but also to the collective processes of adaptation. Keeping the cultural impacts in mind creates solid foundations for the success of change processes and for the humanization of work. Businesses cannot simply not shape activities – even the explicit or implicit renunciation of design measures is a decision! This new perspective is not replacing but complementing the tried and tested practical ways of protecting safety and health at work based upon national standards and regulations. On the basis of five theses below, we elaborate on approaches to the design of the work of tomorrow.

3.2   THESIS 1: WORK SYSTEM MODELS MUST BE EXPANDED AND ADAPTED TO THE PARTICULARITIES OF DIGITAL WORK

Work system models "apply to the design of optimal working conditions with regard to human well-being, safety and health, including the development of existing skills and the acquisition of new ones, while taking into account technological and economic effectiveness and efficiency" (ISO 2016). They therefore have direct relevance to the design of workplace factors that have undergone digital transformation. However, the models must also be adapted, in terms of content and structure, to the change dynamics which are typical for digital transformation.

Most of the models currently in widespread use are composed of similar basic elements that are arranged differently. Based on ISO 6385:2016, the central elements of a work system are:

- Worker: "person performing one or more activities to achieve a goal within a work system".
- Work task: "activity or set of activities required of the worker to achieve an intended outcome". A job is "the organization and sequence in time and space of an individual's work task or the combination of all human performance by one worker within a work system".
- Work equipment: "tools, including hardware and software, machines, vehicles, devices, furniture, installations and other components used in the work system".
- Work environment: "physical, chemical, biological, organizational, social and cultural factors surrounding a worker" (including the workspace).
- Work process: "sequence in time and space of the interaction of workers, work equipment, materials, energy and information within a work system"; work organization stands for work systems interacting to produce a specific output.

In addition, work system models generally address the exchange of *input* and *output* of the work system with their environment.

Based on an understanding of the organization according to systems theory, two model types can be distinguished. (1) A simple closed-loop control model. The basis of this first-order cybernetics is "the idea of monitoring a system, or its targeted control analogous to straight-line cause-and-effect relationships" (Simon 2009, 19). (2) A sociotechnical model, that additionally considers the dynamic interactions between technical and tangible subsystems and social subsystems (Ulich 2013).

Both approaches fall short as a basis for the task of designing work that has undergone digital transformation, as they do not adequately address three important aspects. First, the models do not take appropriate account of the change dynamics. They are based on the assumption that the work system can be organized or corrected once and for all, and can then be run or controlled indefinitely according to defined rules. The nature of the influence of digital transformation, however, is constant and disruptive, which means that there is a permanent need to adapt.

Second, the conceptualization of leadership is based on a traditional understanding of leadership – if it is explicitly taken into account in the organizations at all. In other words, it is assumed that the leader and the experts supporting the leader have all the knowledge and implementation competence necessary in order to design the work system. In a context of constant dynamic change, however, the necessary knowledge of optimal work design is a product of a process of dialog with the active involvement of workers (primacy of participation). Here, the function of leadership is to create a space for participation and thereby establish the best possible conditions for work design.

Third, the models take account of the influence of the corporate culture marginally at most, for example, as an aspect of the work environment. Yet the particular culture of the organization has a crucial bearing on how the technical and social system elements interact and plays a fundamental role in the development of a sustainable culture of prevention in safety and health at work.

3.3   THESIS 2: DIGITAL TRANSFORMATION CREATES TENSIONS WHICH HAVE POSITIVE AND NEGATIVE IMPACTS ON WORK AND HEALTH

The digital transformation presents opportunities as well as risks in terms of the safety and health of people in work systems. Whether a specific technological influence on the work system should be seen as positive or negative depends in each individual case on the particular workplace type, the company, and its culture.

3.3.1   IMPACTS OF DIGITAL TRANSFORMATION ON THE ELEMENTS OF THE WORK SYSTEM

Table 3.1 sets out in concrete terms the central influences of digital transformation on the elements of the work system.

In terms of their effect on the elements of the work system – and therefore on safety and health – the listed changes or influencing factors related to digital transformation can be neutral, positive, negative, or both positive and negative. What is certain is that the interactions reach a degree of complexity that makes it impossible to decide from the outset whether the work-related stressors have critical health-strain consequences. Case-by-case assessments are the only way to shed any light.

TABLE 3.1

Relationship between Work System and Influences Related to Digital Transformation

Element of the work system	Changes and influencing factors related to digital transformation

Worker	• Digital competence is becoming a core competence in all sectors (Holdampf-Wendel 2018)
	• "Just in time learning" is increasingly important (BMAS 2017, 109)
	• Changing values and expectations of work, e.g.,
	  • life plans are becoming more individual and more varied
	  • classic role models are becoming less rigid
	  • many workers want more freedom to improve their private-work-life balance (BMAS 2017, 75)
	  • "more time for the important things", like innovation, social work, leisure (Bauer 2018)

Work task	• The number of activities in which cognitive, information-based, and emotional factors dominate is steadily increasing. In many occupations, there is a shift from the physical demands of the past to predominantly mental demands now (BMAS 2017, 135) • Activities are being automated, not necessarily entire jobs (BMAS 2015) • Job profiles are disappearing, and new ones are emerging (Holdampf-Wendel 2018)
Work process	• Intensification of work: acceleration of the world of work and greater work pressure (Diebig et al. 2018, 57) • Increasing and decreasing autonomy and self-organization • New forms of cooperation and networking: coworking, crowdworking, cocreation (Bauer 2018) • Constant adaptation of work processes to changing value chains (Bauer 2018) • Dynamic integration of customer requirements in work processes (Bauer 2018) • Greater transparency/monitoring/supervision of activities (Diebig et al. 2018, 59) • Management of more complex correlations in processes (Bauer 2018; DGUV 2016, 15–16) • New forms of leadership (Bauer 2018; DGUV 2016, 30 et seq.)
Work equipment	• Artificial intelligence (AI), digital assistance, and tutoring systems (BMAS 2017) • Cobots (collaborative robots) (Bauer 2018) • Systems providing physical support/enhancing abilities (e.g., robots, human-robot collaboration, exoskeletons) • Digital assistance systems (e.g., data glasses, software to support decision-making) • Wearables (e.g., smartwatch, intelligent work clothing)
Work environment	• More flexible working time and location/blurred boundaries (BMAS 2017, 73 et seq.; Diebig et al. 2018, 57; DGUV 2016, 21 et seq.) • Cognitive environments and smartroom technology (ergonomically customizable workplaces – desk height, lighting, acoustics, room temperature, oxygen supply, air humidity) (Bauer 2018)

Autonomy of action/decision-making at work

| High levels of individual responsibility and self-management

Increase in perceived coherence and meaningfulness | | Negative strain from financial/business pressure, increasing dependency on (opaque) processes (organization as a "black box")

Monitoring of behavior/performance with end-to-end data capture

Decrease in perceived coherence and meaningfulness |

Competence development and learning

| Ongoing competence improvement (e.g. in new forms of cooperation and with just in time learning)

Easier learning (e.g. with tutoring systems)

Self-learning algorithms (machine learning) and the parallel analysis of vast quantities of information are enabling AI applications to adapt to people in line with the situation at hand, and to carry out a large number of complex tasks in close collaboration | | Qualification pressure and mental overload

Downgrading (for example because of the declining importance of practical experience or the use of self-learning algorithms/artificial intelligence (AI) to take over problem-solving) |

Occupational safety

| Improved accident prevention with intelligent tools, clothing and equipment (data glasses, gloves, cobots) registering everything that happens around them, allowing a safe physical human-machine collaboration | | Increased accident risk in the event of a system failure due to incorrect responses (with highly complex systems that are poorly understood) |

Participation in work

| New opportunities for participation in work because assistance systems can compensate for physical or sensory impairments (also older workers) | | Exclusion from work due to a lack of flexibility in terms of location, or of people without access to information technology (IT) competences |

Physical burden

| Better protection from physical overload | | Physical complaints due to lack of exercise (e.g. no more changes of location – "sitting is the new smoking") |

Greater flexibility

| A better private-work-life balance thanks to greater flexibility in working time and location, and greater use of opportunities within the organization to increase flexibility | | Possible negative stressors resulting from blurred work boundaries as well as partial absence of defined employment contracts and forms of employment |

Ergonomics

| Better workplace ergonomics, for example with intelligent workplace equipment and smart rooms | | Risk of poor ergonomics at home or in shared workplaces |

Social contacts with new forms of collaboration

| Better networking | | Lack of social integration/belonging or fewer direct social contacts/isolation |

FIGURE 3.1   Opportunities and risks of digital transformation.

3.3.2 OPPORTUNITIES AND RISKS OF DIGITAL TRANSFORMATION

Bearing in mind the constant and disruptive character of digital transformation, the polarizations in Figure 3.1 are currently recognizable.

In addition, polarization is possible between groups of workers because of differing expectations and needs within a workforce regarding the design of working time and location, and tensions may emerge between individual preferences and collective arrangements (BMAS 2017, 80). This list can be used as general guidance when determining the specific influence of digital transformation on a company or on a particular job type. An analysis within the particular organization is essential in order to identify concrete design requirements.

3.3.3 Digression: Example of Determining the Impacts of Digitalization on a Company

By way of example, we present below the qualitative analysis of the impacts of digital transformation in the context of the digitalization initiative of the corporate unit for Environmental Protection, Health Management and Safety (EHS) in Siemens. The goal of the project was to understand the impacts of digital transformation with reference to the three EHS subject areas, in order to identify opportunities and risks as well as action to be taken. In particular, the project aimed to determine whether digital changes entail new psychosocial work stressors and what new requirements will result in terms of occupational safety and health management. The Siemens unit received support in designing the survey and evaluating the interviews from the working group for Applied Medicine and Psychology at Work at the Institute for Occupational, Social and Environmental Medicine of the University of Munich.

The analysis was conducted by means of semi-standardized interviews with 52 leaders from 15 countries and 16 different job types. The interviews were structured according to the elements of the work system and their content was based on a comprehensive survey of the literature regarding the influences of digital transformation on work. In total, the evaluation considered about 1,500 individual statements (there were some multiple assignments of statements to work system elements). The main results of the survey are as follows:

- The change in working caused by digitalization, and especially the influence on health and wellbeing, was rated as positive overall by 61% of interviewees, as negative or potentially risky by 31%, and as neutral by 8%.
- The main impacts of digital transformation are seen in relation to the work task followed by work equipment and work environment; the influence on work organization and the workers was rated much lower.
- For work tasks, the main opportunities (positive effects) are felt to be the variety and coherence as well as the technical support (work equipment); the main risks are felt to relate to the complexity and intensity of work.
- In terms of work equipment, the surveyed leaders expect digitalization to make much better use of resources thanks to IT applications; they are more critical of the possibly inadequate ergonomics and adaptation of software to human abilities and needs.

- The leaders expect the digital changes to make it easier to organize work; meanwhile they see risks around social integration of workers and the blurring of boundaries in tasks and roles.

The results of this survey confirm what the literature has found: that the influences of digital transformation are always seen as a balancing act between opportunities and risks with regard to safety and health at work. It is also plain that the estimation of the impacts of digitalization can be very different even within the same company. Of particular note are the largely positive expectations of the benefits of digital transformation for work, and the very specific suggestions regarding the design and support requirements from the EHS organization in the company. For example, the interviews suggest new aspects or emphases of investigation when carrying out risk assessments of workplaces and tasks that have undergone digital transformation, such as the ergonomics of new IT-based work equipment or social support in the context of new forms of collaboration.

3.4 THESIS 3: HEALTH AND SAFETY RISK ASSESSMENTS ENABLE THE DESIGN OF WORK THAT HAS UNDERGONE DIGITAL TRANSFORMATION

Risk assessment is a process designed to prevent occupational safety and health risks. In work systems affected by digital transformation, the emphasis shifts to the assessment of mental stress. This opens up the opportunity to identify and harness resources over and above the risk-oriented approach which is generally taken at present (Borg 2017; Borg et al. 2018). The risk assessment covers the entire work system. In the digital world of work in particular, specific procedural concepts, based on systemic thinking, must be developed for the holistic risk assessment that has been required for years in the work sciences. This requires greater dynamism in the procedure, achieved by altering the intensity, the methodology, and the prospective reach. Relevance is ideally guaranteed with iterative processes based on participation. Relationships between opportunities and risks, stressors and resources can be evaluated in this way, culminating in continuously evaluated measures. One challenge is the integrated assessment of different physical and mental stressors (GDA 2014).

A risk assessment of work-related mental stress is a legal requirement in many countries (SLIC 2012; e.g., BMAS 2013 for Germany). The effect of work on people is usually described by the Stress-Strain Concept (Rohmert and Rutefranz 1975). *Work-related mental stress* is defined as the "total of all assessable influences impinging upon a human being from external sources and affecting it mentally" (ISO 2017). *Stress* is therefore *any* form of influence of work which can create an effect in the human organism. *Strain* is the "immediate effect of mental stress within the individual" (ibid.). Stress should be understood as value-neutral. It can have positive impacts (consequences of mental strain) such as stimulating and coaching effects, as well as negative impacts such as a feeling of monotony, and it can contribute to mental exhaustion and stress responses, or to the development of mental disorders over the long term (ibid.).

The Stress-Strain Concept is used in an assessment of working conditions, in particular to analyze the mental work stress. However, it is better suited to the retrospective assessment of working conditions, as it is based on defined, scientifically confirmed stressors. Yet digital transformation is characterized by a large number of influencing factors which are constantly changing and reconfiguring, and which cannot be identified fully or in advance. The question is how to find assessment and design processes that can help to humanize work even though the configuration of stressors is just beginning to emerge.

One option is to examine the current interactions between people and the work system. This means that an assessment of work in which the strain consequences are as yet unknown or unexplored is only possible for a specific period in specific workplaces. The participation of those affected is a good opportunity to make up for the lack of empirical knowledge in the work sciences, while increasingly empowering workers by giving them responsibility to shape their own world of work.

With reference to digital transformation, the consequences for the risk assessment are as follows:

- New stressors caused by digital transformation must be taken into account in the content lists of the risk assessment (Diebig et al. 2018). This primarily applies to psychosocial and ergonomic stressors. The risk assessment of mental stress is becoming vitally important for knowledge workplaces.
- With the health sciences unable to give us reliable insights into the new "digital" forms of stress, participation in the risk assessment process is a way of identifying relevant negative stressors and developing practical countermeasures. Participation goes hand in hand with a different understanding of leadership and a change of culture in the organization. In this context, increasing account must also be taken of the international make-up of workers (Diebig et al 2018; see Yorio et al., Chapter 6 in this book).
- ISO 6385 (2016) and part 2 of ISO 10075-2 (1996) (ISO 2017a) state that technical, organizational, and human factors, and the interactions between them, must be considered in the design of work systems. This inclusion in the context of a holistic risk assessment of working conditions must be pinned down conceptually and methodologically.
- The procedures or instruments used for risk assessment must cover all work locations in which workers are active in work systems that have undergone digital transformation (Diebig et al. 2018).
- The risk assessment process must be made more dynamic. The common practice of ascertaining mental stress at two- or three-year intervals no longer appears to be appropriate bearing in mind the accelerating changes taking place in work systems undergoing digital transformation (Diebig et al. 2018). In future, the working conditions will ideally be analyzed and designed at any time, even in real time, through the use of new information sources from intelligent work environments, with those involved given greater autonomy and design competency.

3.5 THESIS 4: WORK SYSTEMS THAT ARE AFFECTED BY DIGITAL TRANSFORMATION REQUIRE A REFLEXIVE CONCEPT OF LEADERSHIP

Here, leadership is considered as a function of the organization, independently of the individuals. The purpose of leadership is to pave the way for decisions and to balance the health resources and risks in a world of work undergoing digital transformation. This involves observing its own actions and opening them up for discussion. This reflexive attitude forms the basis of the design competency of leadership in an organization.

Traditional work system models tend to neglect the influence of leadership on how the system elements relate to each other. In fact, it can be assumed that leadership plays an important role as intermediary between the technical and social subsystems, and is therefore a key lever in the design of work affected by digital transformation (Figure 3.2).

A question arises about the function and form of leadership in increasingly agile, self-organizing, and networked organizations (see Brendebach, Chapter 4 in this book), namely: do we still need leadership, and if so, what kind?

Leadership has traditionally been linked to the attributes and capabilities of individuals (Neuberger 2002; see Schöbel, Chapter 5 in this book). From a systemic point of view, though, leadership should be understood as an attribute and capability of an organization. As a social system, the organization develops a form of leadership which is conditioned by its historical conflicts, in order to maintain its problem-solving capacity into the future (Wimmer 2009). Leadership is therefore the attempt of an organization to increase the likelihood that the right decisions are

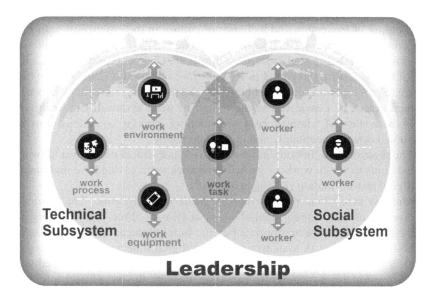

FIGURE 3.2 Leadership and the dynamically changing elements of the work system.

taken for its own survival. The outcomes can be very different in terms of organization designs and strategies, each of them aimed at controlling imponderables and guaranteeing the long-term existence of the organization. For example, the exchange of necessary resources with the environment – depending on resource dependency and availability – can yield organizational solutions that encompass both internal (buffering, greater flexibility, etc.) and external programs (cooperation, contractual arrangements, etc.) (Schreyögg 2003, 89). The digital transformation exerts internal and external adaptation pressure, and as such it has a varying impact on the organization and therefore on the dynamics of the organizational change process, ultimately influencing the cultural diversity of leadership.

Based on a modern understanding of leadership, Wimmer (2009) proposes seeing the primary task of leadership today as finding and implementing organizational solutions that match the dynamic development of the relevant environment. In his opinion, power asymmetries continue to be necessary in order to make decisions and implement them. Even so, Wimmer favors a vision of "leadership which primarily guides the attention of workers *no longer upwards* – this is no longer the driving force in day-to-day activities – *but outwards.* What does the customer need? What is needed for successful interaction in the various networks which span organizational boundaries?" (Wimmer 2009, 25; italics by the authors).

Aside from the management of internal and external relationships, leadership in the world of work undergoing digital transformation faces the central challenge of mediating decisions about the design of the work system which allow coordinated action to be taken. This appears to be difficult because complex problems are increasingly associated with diverging interests and perspectives which must be taken into account. Success is only possible if workers are able to actively participate in the development and design process – in other words developing, discussing, and ratifying in a discursive process (Hartmann 1995). This can work particularly well in communication processes that result in ongoing "sense-making" in the organization (Wimmer 2009, 30). The important thing is not whether the communication is based on analog or digital media or is designed using AI programs. Instead, it is crucial that the medium and the content are closely aligned. In other words, communication structures that are able to cope with complexity and suitable media guarantee the flow of information and support decision-making. Meanwhile, a flow of information which is up to the task is also a resource maintaining the health and performance of workers.

Leadership is always exercised through communication. All workers constantly monitor whether the communicated content matches the actions really taking place. Leadership therefore influences the culture in every direction, and on the negative side, a lack of fairness or integrity can cause frustration and resignation among workers, becoming a health risk in the long term and resulting in reduced performance. Or on the positive side, leadership can be a health resource in the work system through respect, credibility, and participation. To make the ethical foundations of the leadership style more transparent, culture development programs are a good way of supporting organizations as they affirm the premises of their actions.

The DGUV (German Social Accident Insurance) has published guidance on developing leadership principles (DGUV 2019a). It contains support for the ethics-based analysis and reflection of the leadership style. It describes proactive, value-creating premises for action in the development of a culture of prevention, in which leaders should:

(1) Take responsibility: make decisions based on the values of respect for others, fairness, legality, and integrity, and be accountable for their own decisions.
(2) Give meaning, agree goals: provide guidance and agree realistic goals jointly with the workers in a transparent process, hold regular feedback meetings, and support the workers' development goals.
(3) Involve workers: demonstrate cooperative, worker-oriented, and collaborative leadership behavior in which participatory discussion forums are deployed as self-evident leadership instruments.
(4) Provide support: harmonize competencies, resources, and tasks, allow for freedom of action, design health-promoting working conditions, and actively support workers.
(5) Maintain contact: take time for each other, express recognition and appreciation.
(6) Develop leadership competence: reflect on their own leadership behavior, be open to feedback, and take up training opportunities.
(7) Apply healthy leadership to themselves: pay attention to their own health and be clear that their own behavior is taken as a model for others.

The leadership principles can be used as normative contributions to the internal dialog, as a way of provoking reflection about the prevailing leadership style in the organization. First, this allows leadership to be considered as a function, independently of the individual concerned. And second, it stimulates constructive debate about the contribution of leadership to the design of internal and external relationships and to the patterns of decision-making that are able to cope with the development dynamics.

3.6 THESIS 5: DIGITAL CHANGE IS MANAGED MORE EFFECTIVELY THROUGH THE SYSTEMATIC PROMOTION OF A CULTURE OF PREVENTION

The quality of interaction between technical and social elements of the work system is decisively shaped by the particular corporate culture. The cultural aspect makes a significant contribution to resolving the conflict between financial and strategic goals and the requirements of humane work in organizations undergoing digital transformation. Furthermore, a culture of prevention significantly improves the integration between occupational safety and occupational health management.

3.6.1 CULTURE OF PREVENTION

Each organization has its own characteristic and unique corporate culture – there is no such thing as no culture. It is dependent upon its technological core (Thompson 2017), in other words, the structures and practices serving to transfer the input into the desired output, for example, the production of fridges in the electrical industry or power generation in the energy sector. Then there are the characteristics of the world outside the organization, which could be stable or competitive, for example, and exert more or less pressure to change. This means that culture is a genuine attribute of organizations and simultaneously the result of a long-lasting process of social interaction and mutual adaptation in the organization.

Edgar Schein describes culture as

> (t)he pattern of basic assumptions that a given group has invented, discovered, or developed in learning to cope with its problems of external adaptation and internal integration, and that have worked well enough to be considered valid, and therefore, to be taught to new members as the correct way to perceive, think, and feel in relation to these problems.

(Schein 1992, 6)

So, from a systemic point of view, a significant portion of the workers' everyday actions can be traced back to the organization's cultural norms. Companies are social systems which demarcate themselves from their environment through the formal affiliation of their members and which develop certain patterns of describing dangers and risks and dealing with perceived problems (Luhmann 2017). The differences in dealing with risks are primarily a result of an organizational learning process which speeds up when under pressure to change and slows down as soon as the environment exerts less pressure to adapt.

After entering the organization, workers acquire the competence to participate in communication. They may well prefer to see themselves as autonomous agents, making decisions on the basis of their own rational thought processes. They are rarely aware, though, that their everyday actions in the context of the organization are an unfolding of predetermined patterns established in the corporate culture and absorbed during the process of corporate socialization (Elbe 2016, 62). In other words, their behavior, which they are likely to explain in terms of their own rational decisions or even momentary emotions, are substantially determined by a corporate normality which is derived from the history of the company and which dominates their decision behavior. At the same time, everyone in work systems is exposed to individual influences from digital transformation (for example, from private technical interests or influences from the private environment) and introduces them into the work context. Accordingly, corporate culture is a collective phenomenon which influences and is also influenced by the limited number of people with organizational affiliation. The approach to occupational safety risks can therefore be described as a collectively accepted interpretation of reality which arises through exchange with the environment and during daily activities, and which has a lasting and visible influence on the accidents that happen and on health.

This also includes the forms and behaviors in the corporate culture that have "somehow" become embedded in over time, affecting how a company deals with health risks. People see that some behavior or other worked well in day-to-day activities, so it is repeated. Only when these unwritten rules are broken (for example by outsiders) they suddenly become visible. Britta Schmitt-Howe (2018) has developed a typology describing three basic assumptions that can be observed when safety and health risks are handled and monitored:

- Type A is focused on the constant development of operational prevention: in occupational safety, if you stop getting better, you stop being good.
- Type B is concentrated only on the most important, known risks: if you make sure that machinery and equipment are in good order, you have already done most of the work around safety and health – a rather fatalistic attitude.
- And type C relies on the competencies of workers. This attitude is based on the conviction that qualifications and creativity are the best defense against health risks and accidents. It is extremely problematic because if the dominant assumption in a company is that risks can only be monitored at the individual level, it follows that steps like the risk assessment are perceived as irrelevant formalities. After all, "everyone can and must look out for themselves" (ibid.).

Type A companies focused on a participatory process of continuous improvement are more willing than the other types to openly discuss stressors affecting the health of workers in organizations undergoing digital transformation and to develop concrete measures to deal with them. The number and quality of the measures, as well as the commitment with which organizational and structural implementation is driven forward, seem to depend directly on the basic assumptions which are typical for these corporate cultures. It may appear questionable whether in fact there are only few typical corporate cultures for handling safety and health risks. Yet even if a greater cultural variability must be assumed, the relationship between cultural influence and organizational action seems to be plausible and relevant to culture development processes.

This throws up the issue of how the development of a culture of prevention can be encouraged in the context of, and in order to design, the influences of digital transformation. "The term 'culture of prevention' describes both a broader understanding of prevention and a new level of quality for preventive action; safety and health are integrated into all activities and thus, become a self-evident, lived part of our professional and daily lives" (Bollmann 2018, 122). In principle, this is a long-term organizational development process. It requires a crucial new element – continuous self-observation of the culture of the organization – and its success depends on new, useful patterns of behavior being perceived as successful in everyday experience.

To give the culture development a direction, a challenging but attractive goal should be set. A Vision Zero (Zwetsloot et al. 2017) has taken hold for the development of a culture of prevention: zero accidents, good health, sustained promotion of the wellbeing of workers. In this way, hard strategic and financial goals, such as fewer accidents and days off, high productivity, and compliance with quality standards, can be pursued alongside soft goals which arise from ethical issues or which help to improve or preserve a positive corporate image.

The Zero Harm Culture @ Siemens (ZHC) program can be named as an example, which aims to boost safe behavior among in-house employees as well as contractors.

The ZHC master plan, the implementation toolbox, and the corporate communication address the two central dimensions of the safety culture, giving them equal weight and linking them closely:

- Objective, system-related aspects of culture, such as roles and formal responsibilities, risk assessment, investigation of incidents, training, and safety management of external companies.
- Subjective, person-related aspects of culture, such as belief in a zero-accident culture, leadership commitment, responsibility for safety, communication, and recognition.

Globally, between 2012 and 2018, the accident figures fell by 40% at Siemens and the number of fatal accidents (mostly contractors) fell by 80%, in large part thanks to Zero Harm Culture @ Siemens.

3.6.2 DIGRESSION: THE KOMMMITMENSCH CAMPAIGN AS A CALL FOR CULTURE DEVELOPMENT

Alongside proved technical and organizational measures, the German Social Accident Insurance Institutions for the industrial and the public sector are using the *kommmitmensch* campaign (literally translated as "come with [us], man"). It aims to facilitate humanization of work and to permanently embed safety and health in corporate cultures thereby reducing the frequency of accidents. Here, "culture of prevention" is understood normatively. The core message is as follows: "Safety and health are values for all people, for each organization and for society. They should become the focus of every action. Preventive action is rewarding and meaningful" (DGUV 2019b). Ways to influence culture development are seen in these areas of activity: leadership, communication, participation, error culture, and working atmosphere, and also in the inclusion of prevention as an integral part of all tasks and decisions. For all areas of activity there is a concrete description of how they can advance a culture of prevention:

- Safety and health are important to the leadership of the business: the leadership defends the necessary time and resources and acts as a role model itself.
- Targeted and respectful communication within the company means that people communicate on an equal footing and all workers receive the information they need for their particular work.
- Timely and ongoing involvement of workers in operational decisions signals appreciation and increases acceptance of all safety and health measures in the company.
- Errors leading to illness or accidents at work should be avoided if possible. This requires targeted and open discussion of errors and near-misses. If something does happen, the incident should be addressed in order to learn from it. This is the lived error learning culture.

- The working atmosphere is characterized by the degree of mutual support all members of the organization give each other. Appreciation and fairness ensure that they feel at ease.
- Safety and health must take place through systematic integration into all operational tasks. That means taking account of safety and health in all decisions and actions in the company.

The areas of activity are closely linked. In their interaction, they aim to build up a resilience to internal and external destabilizing influences. As the organization grapples with the issues which arise, it learns to apply participative techniques in decision-making processes, helping to detect blind spots and to cast a critical eye on its own expectations. For example, the usefulness of embedded routines can be examined at the crucial moment, and full account can be taken of safety and health (Gebauer 2017, 6).

The German Social Accident Insurance Institution for the Energy, Textile, Electrical and Media products sectors (BG ETEM) recommends that companies establish or develop a culture of prevention in a targeted way by means of a systematic organizational development process. First, a vision is developed in the context of an internal dialog, providing the impetus to move things in the right direction with energy and commitment. The vision is then integrated into the company's strategic goals and embedded in guiding principles and rules. Credibility and meaningfulness are very important here. As the process continues, it must be possible to convey to each worker and each leader what is expected, where the path is leading, and why we are following it together. It must be clear to everyone involved at all times exactly how much progress has been made towards the development of a culture of prevention. To actively support the development process, prevention culture consultants from BG ETEM encourage the creation of a process architecture that uses dialog-based methods to promote continuous self-observation and reflection. Active participation of leaders and workers is an essential element of the concept. The leadership has a central part to play – it ensures that attention is focused on the development of the new culture of prevention.

3.7 CONCLUSION

The impacts of the change dynamics in digital transformation make it necessary to continuously design work systems. Work system models based on ISO standards provide a meaningful and useful framework. However, in light of a modern understanding of organizations, for future worlds of work they must be expanded to include a systemic view of leadership. In the methodologies, too, greater account must also be taken of the culturally determined dynamics of the elements of work systems as they interact.

In practical implementation, the procedural logic of the risk assessment suggests itself. The evaluation of the rapidly changing influences on safety and health of workers is less and less able to fall back on design insights from research in the work sciences. In each individual case, the impacts of the digital transformation are somewhere between the extremes of "opportunity" and "risk". Where the pendulum swings is determined in the particular situation. This shines a light on the competences of the organization to design the processes of change.

Leadership as a central function of the organization guides worker participation, of a kind which is methodologically appropriate for the change dynamics, towards identifying stressors and resources in the work. Leadership must ensure that the organization always has at its disposal processes and resources to develop safety and health measures with the right time span. The technical possibilities of digitalization are proving to be an opportunity for the integrated recording of physical and mental stressors.

Systematic reflection about actions and decisions in the organization augments the quality not only of leadership but also of communication, the error learning culture, and the integration of safety and health into business processes. Leadership thereby firmly ties business goals to the mission to humanize organizations and make them future-proof. The result is a culture of prevention – in other words, safe and healthy work in a successful company.

REFERENCES

Bauer, W. 2018. *Zukunftsräume Schaffen! Neue Perspektiven für Arbeit*. Speech at Fraunhofer IAO-Future Forum, 2 February 2018, Stuttgart.

BMAS – Federal Ministry of Labor and Social Affairs. 2013. *Act on the Implementation of Measures of Occupational Safety and Health to Encourage Improvements in the Safety and Health Protection of Workers at Work (ArbSchG)*. Berlin: Beck.

BMAS – Federal Ministry of Labor and Social Affairs. 2015. *Übertragung der Studie von Frey / Osborne (2013) auf Deutschland – Endbericht*. Forschungsbericht 455. Berlin: Beck. https://www.bmas.de/DE/Service/Medien/Publikationen/Forschungsberichte/Forschungsberichte-Arbeitsmarkt/forschungsbericht-fb-455.html (accessed December 12, 2019).

BMAS – Federal Ministry of Labor and Social Affairs. 2017. *White Paper Work 4.0*. Berlin: Beck. https://www.bmas.de/EN/Services/Publications/a883-white-paper.html (accessed December 12, 2019).

Bollmann, U. 2018. Competences for a culture of prevention – Conditions for learning and change in SMEs. In: G. Boustras and F. Guldenmund, eds. *Safety Management in Small and Medium Sized Enterprises (SMEs)*, 121–141. Boca Raton: CRC Press / Taylor & Francis.

Borg, A. 2017. *Anwendung von Modellen der Sicherheitskultur im betrieblichen Kontext und deren Übertragbarkeit auf Präventionskultur*. Bexbach: CBM.

Borg, A., C. Digmayer, J. Reinartz, and E. M. Jakobs. 2018. Sicherheitskultur: Wegbereiter für Digitalisierung. In: R. Trimpop, J. Kampe, M. Bald, I. Seliger, and G. Effenberger, eds. *Psychologie der Arbeitssicherheit und Gesundheit. Voneinander Lernen und miteinander die Zukunft Gestalten! 20. Workshop 2018*, 231–234. Kröning: Asanger.

DGUV – Deutsche Gesetzliche Unfallversicherung e.V. 2016. *Neue Formen der Arbeit. Neue Formen der Prävention. Arbeitswelt 4.0: Chancen und Herausforderungen*. Berlin: DGUV. https://publikationen.dguv.de/dguv/pdf/10002/dguv-nfda_de_barrierefrei.pdf (accessed December 12, 2019).

DGUV – Deutsche Gesetzliche Unfallversicherung e.V. 2019a. *Führung – Führungsleitlinien Erstellen und umsetzen*. Hürth: CW Haarfeld.

DGUV – Deutsche Gesetzliche Unfallversicherung e.V. 2019b. *Kommmitmensch campaign*. https://www.kommmitmensch.de/die-kampagne/uk-und-bg/ (accessed December 12, 2019).

Diebig, M., F. Jungmann, A. Müller, and I. C. Wulf. 2018. Inhalts- und prozessbezogene Anforderungen an die Gefährdungsbeurteilung psychischer Belastung im Kontext Industrie 4.0. *Zeitschrift für Arbeits- und Organisationspsychologie* 62(2), 53–67.

Elbe, M. 2016. *Sozialpsychologie der Organisation: Verhalten und Intervention in sozialen Systemen.* Wiesbaden: Springer.

GDA – Mental Health Working Program. 2014. *Recommendations of the Institutions of the Joint German Occupational Safety and Health Strategy (GDA) for Implementing Psychosocial Risk Assessment.* Berlin: Management of the GDA Mental Health Working Program, c/o Federal Ministry of Labor and Social Affairs.

Gebauer, A. 2017. *Kollektive Achtsamkeit organisieren. Strategien und Werkzeuge für eine proaktive Risikokultur.* Stuttgart: Schaeffer-Poeschel.

Hartmann, A. 1995. "Ganzheitliche IT-Sicherheit": Ein neues Konzept als Antwort auf ethische und Soziale Fragen im Zuge der Internationalisierung von IT-Sicherheit. In: Bundesamt für Sicherheit in der Informationstechnik, ed. *Fachvorträge 4. Deutscher IT-Sicherheitskongress. Sektion 7, BSI. 7165,* 1–13. Bonn: BSI.

Holdampf-Wendel, A. 2018. Berufe für die Zukunft. Speech at Fraunhofer IAO-Future Forum, 2 February 2018, Stuttgart.

ISO – International Organization for Standardization. 2016. ISO 6385 – Ergonomics Principles in the Design of Work Systems. https://www.iso.org/obp/ui/#iso:std:iso:6385:ed-3:v1:en:term:2.2 (accessed December 12, 2019).

ISO – International Organization for Standardization. 2017. ISO 10075-1: Ergonomic Principles Related to Mental Workload. Part 1: General Issues and Concepts, Terms and Definitions. https://www.iso.org/obp/ui/#iso:std:iso:10075:-1:ed-1:v1:en (accessed December 12, 2019).

ISO – International Organization for Standardization. 2017a. ISO 10075-2:1996: Ergonomic Principles Related to Mental Workload – Part 2: Design principles. https://www.iso.org/obp/ui/#iso:std:iso:10075:-2:ed-1:v1:en (accessed December 12, 2019).

Luhmann, N. [1991] 2017. *Risk: A Sociological Theory.* London: Routledge / Taylor & Francis.

Neuberger, O. 2002. *Führen und führen lassen: Ansätze, Ergebnisse und Kritik der Führungsforschung.* Stuttgart: UTB.

Rohmert, W., and J. Rutenfranz. 1975. *Arbeitswissenschaftliche Beurteilung der Belastung und Beanspruchung an unterschiedlichen Industriearbeitsplätzen.* Bonn: Bundesministerium für Arbeit und Sozialordnung.

Schein, E. [1985] 1992. *Organizational Culture and Leadership. A Dynamic View.* 3rd ed. San Francisco: Jossey-Bass.

Schmitt-Howe, B. 2018. What kind of prevention cultures are prevailing? Typical dialogues on occupational safety and health in German companies. In: A. Bernatik, L. Kocurkova, and K. Joergensen, eds. *Prevention of Accidents at Work,* 171–177. London: CRC Press / Taylor & Francis.

Schreyögg, G. 2003. *Organisation. Grundlagen Moderner Organisationsgestaltung.* 4th ed. Wiesbaden: Gabler.

Simon, F. 2009. *Einführung in Systemtheorie und Konstruktivismus.* 4th ed. Heidelberg: Carl Auer.

SLIC – Senior Labor Inspectorate Committees. 2012. *Psychosocial Risk Assessments – SLIC Inspection Campaign 2012, Final Report.* Luxembourg: European Commission. http://www.sevosh.org.gr/el/Images/News/SLIC-2012/SLIC1 (accessed December 12, 2019).

Thompson, J. D. [1967] 2017. *Organizations in Action: Social Science Bases of Administrative Theory.* New York: Routledge / Taylor & Francis.

Ulich, E. 2013. Arbeitssysteme als Soziotechnische Systeme – Eine Erinnerung. *Psychologie des Alltagshandelns* 6(1), 4–14.

Wimmer, R. 2009. Führung und Organisation – Zwei Seiten ein und derselben Medaille. *Revue für Postheroisches Management* 4, 20–33.

Zwetsloot, G., S. Leka, and P. Kines. 2017. Vision Zero: From accident prevention to the promotion of health, safety and wellbeing at work. *Policy and Practice in Health and Safety* 15(1), 1–13.

4 Competence Management

Between Command and Control, Self-Organization, and Agility

Florian Brendebach

CONTENTS

4.1 INTRODUCTION

If competence is understood as a "disposition for self-organization" (Erpenbeck 2013, 308), it becomes possible to use competences as a source of autonomous and creative action in open-ended contexts (Arnold 2010, 172; Erpenbeck 2013, 314).

Competence management refers to an organization's efforts to influence these dispositions in a systematic manner. A top-down conception of competence management suits the organization's governance needs and provides a high degree of guidance, but in an increasingly complex, dynamic, and unpredictable world – as expressed in the popular term "VUCA world" (Millar et al. 2018) – such a conception is reaching its limits. There is a growing need to shift the focus from the perspective of anticipation to the perspective of responsiveness. It is in this context that the idea of agility has taken hold, and competence management must become agile as well, eliminating the boundaries between learning and work (Sauter et al. 2018, 67).

The policy field "occupational safety and health" (OSH) encompasses binding legislation and the combined efforts of government and nongovernment actors to promote a "new culture of health in the workplace" (BMAS 2013; 2018) and a "culture of prevention" (DGUV) that integrates aspects of risk prevention and health promotion into a comprehensive perspective covering all stages of life (World Health Organization (WHO) 1986; Bollmann 2018, 122). From this perspective, competence management within an organization provides a context for employees to act and experience their work and, ultimately, a context for their safety and health. At the same time, it is a framework model for the systematic development of safety- and health-related competences.

Based on key elements of a conception of competence and competence development (4.2), the following chapter will contrast a top-down approach to competence management with an agile model of competence management, which includes a discussion of each from the perspective of occupational safety and health at the end of the respective subchapter (4.3).

4.2 BASIC CONCEPTS

This section provides an introduction to some basic conceptual principles, beginning with a definition of the term competence as it is used in this chapter (4.2.1), followed by a presentation of key aspects of competence development (4.2.2) and an overview of relevant organizational concepts (4.2.3).

4.2.1 THE CONCEPT OF COMPETENCE

The concept of competence has grown sharply in importance in recent decades. Even though not universally accepted (Becker 2013, 5–6), it has become a household term in discourses about general education, vocational training, and lifelong learning. And yet there is no standard definition, and users of the term tend to emphasize different aspects (Klieme et al. 2007, 5; Erpenbeck et al. 2017, XXI–XXIV).

It is helpful to distinguish the concept of competence from the concept of qualification. A qualification refers to clearly defined third-party requirements – with previously known solutions. Competence, by contrast, refers to finding self-directed solutions for problems in open-ended situations in which the steps towards a solution cannot be identified in advance (Arnold 2000, 269; Erpenbeck 2013, 313–314). In contrast to an understanding of competence that is limited to cognitive aspects and serves as the basis of international student assessment studies (PISA, TIMSS) (Klieme et al. 2007; Erpenbeck et al. 2017, XXII–XXIII), competence understood as a disposition for self-organization calls for the comprehensive integration of all dimensions that regulate human actions:

> Competences are dispositions for the self-organization of mental and physical actions if dispositions are understood to mean the totality of inner prerequisites for the mental regulation of activities developed prior to a specific moment of action.

(Erpenbeck 2013, 308, author's translation)

To be able to act self-directed in open-ended situations particularly requires internalized criteria for orientation:

> What all conceptions [of this notion of competence] have in common is the development of a subjective potential for self-directed action in various domains of society. This subjective capacity for action is not only tied to the acquisition of knowledge; it also includes the adoption (…) of criteria for orientation as well as personal growth.

(Arnold 2010, 172, author's translation)

Some examples: the ability to create a company balance sheet is a qualification. It is not a trivial task, but how it is done can be explained in advance. By contrast, sitting down with an employee to discuss a co-worker's potential addiction problems requires competence. It includes cognitive as well as emotional and motivational aspects, and no precise description of the steps towards accomplishing this task is available in advance; the best one can do is create a rough outline. Likewise, the ability to use software to coordinate appointments within a project is a qualification, whereas the ability to coordinate a project (e.g., the introduction of an occupational safety and health system) must be considered a competence.

4.2.2 ASPECTS OF COMPETENCE DEVELOPMENT

Closely tied to the competence debate are various concepts of self-organization. Since the 1960s, a wide range of concepts emerged that were essential in shaping today's understanding of the term. These include but are not limited to the concept of autopoiesis by Humberto Maturana and Francisco Varela, Hermann Haken's synergetics, the concept of dissipative systems according to Ilya Prigogine, or Manfred Eigen's research on molecular evolution (Heuser 2016, 155). What connects these theories is the insight that the behavior of systems and organisms is not based on the objective qualities of an environmental stimulus but rather on the specific ways in which this stimulus is processed within the system. This means the possibility of controlling systems from the outside is limited (Krohn and Küppers 1990, 11). In education, this thought was introduced and discussed primarily through Maturana and Varela's theory of autopoiesis and structural determination:

> A state-determined system is a dynamic system in which changes to that state (…) are determined by the structure of the system and not by the impact of an agent independent of that structure. Every state-determined system, through its organization and its structure, defines the realm of possible interactions that may trigger changes in its internal state.

(Maturana 1985, 278, author's translation)

For a self-organized organism, therefore, environmental stimuli, whether they are learning requirements or incentives for action, merely represent impulses that, rather than producing a linear effect, are processed on the basis of the organism's own structure.

The concept of self-organization is not identical to the concept of self-directed learning. The prevailing definition of *self-directed learning* is that offered by Malcom Knowles:

> In its broadest meaning, "self-directed learning" describes a process in which individuals take the initiative, with or without the help of others, in diagnosing their learning needs, formulating learning goals, identifying human and material resources for learning, choosing and implementing appropriate learning strategies, and evaluating learning outcomes.

(Knowles 1975, 18)

From a systemic-constructivist perspective, by contrast, learning and the acquisition of competence always take place in a *self-organized* manner on the basis of individual structures, with learning goals defined either by the individual or by others. That said, self-directed learning (often) does ensure a strong alignment of internal structure and the selected learning goals.

At the inter-individual level, incentives to acquire competences find different preconditions: the internal structures of employees. In this context, it is helpful to adopt a distinction made by Danish scholar Knut Illeris in his extension of the concepts of Jean Piaget (1896–1980). Illeris (2006, 2010) distinguishes four types of learning that may refer to knowledge as well as to motivation and values. (1) *Cumulative learning* takes place if a (cognitive) scheme or an (affective) pattern is formed for the first time without being linked to a pre-existing scheme or pattern. (2) *Assimilative learning* occurs if a new element is linked as an addition to a scheme or pattern that is already established. (3) *Accommodative learning* involves situations in which a new element forces the learner to modify an existing scheme or pattern. (4) Finally, *transformative learning* refers to the process of a fundamental, comprehensive transformation of an organism's schemes or patterns, a type of transformation much like a personality change.

Relating Illeris's four categories to the dimension of directedness with the two poles of other-directedness and self-directedness results in a four-field scheme (Figure 4.1) that illustrates distinct contexts of competence acquisition (Pätzold and Brendebach 2018):

The more learners need to change their own schemes and patterns, the more energy they need to accomplish learning (Illeris 2006, 34; Pätzold 2011, 18–19). At the same time, a higher need for change often goes hand in hand with a higher degree of uncertainty (Illeris 2006, 34; Pätzold and Brendebach 2018). If a learning requirement challenges fundamental schemes and patterns in their self-perception, people (often) respond by activating defense mechanisms (Illeris 2006, 35). Mastering a learning requirement that is deemed to be important can often be experienced as relief or relaxation (ibid., 34). Finally, increases in the degree of perceived other-directedness may involve a growing risk of (learning) resistance.

In terms of teaching methodology, it is possible to identify two poles with specific concepts: other-directed assimilation usually means that a qualification is required. Following a logic of instruction, clearly defined knowledge that builds on existing schemes and patterns is communicated in a linear manner. Typical formats include a classroom presentation or safety instructions at the workplace.

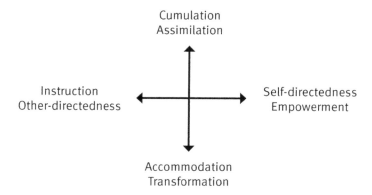

FIGURE 4.1   Contexts of competence acquisition.

If the learning content can be linked to existing schemes and patterns, then this is the point where the implementation of a legal standard can be understood as an other-directed assimilation. For example, if an employer implements the German Noise and Vibration Health and Safety Regulation (LärmVibrationsArbSchV) on the basis of the German Occupational Safety and Health Act (ArbSchG) and the corresponding EU directive, this constitutes a case of other-directed assimilation as far as employees already acknowledge the general importance of noise-reduction. If the employer chooses to act proactively early on in the planning stages by taking measures to avoid the production of noise in the first place and by getting employees involved in that process, the aspects of initial other-directedness and subsequent participatory self-directedness overlap. If employees eventually take the initiative to reduce noise exposure at the workplace, this process fully leads to self-directed assimilation. The organizational management style that corresponds to the logic of instruction is "command and control".

In the case of complex competences, however, such as healthy leadership competence (Hahnzog 2015) or agile teamwork competence, the logic of instruction will no longer produce the desired outcomes. For the most part, complex competences (also) encompass aspects of accommodation (Illeris 2006, 34) and must be acquired in a self-directed manner in suitable contexts. This must be understood as the pole opposed to other-directed assimilation. Competence acquisition must be "facilitated" as self-directed learning. The concept of "teaching as facilitated learning" relates to the design of learning situations that enable employees to acquire competences on the basis of their own internal structure. In such a setting, the boundaries between learning and action are blurred, and the teacher becomes a learning facilitator, offering input to learners without falling prey to the illusion of a linear transmission of knowledge (Arnold and Gómez Tutor 2007; Sauter et al. 2018). This calls for a twofold shift in perspective: (1) away from the illusion of objective knowledge and towards a focus on learners' specific schemes and structures and (2) away from a linear creation of ability towards the self-directed and active acquisition of competence (see Reitz and Schröder, Chapter 11 in this book). A "facilitated" learning situation may be an isolated work situation, but complex work processes, such as participating in a project team, may also be interpreted and designed along those lines. In such a setting, the focus of organizational management moves away from "command and control" towards the mutual coordination of needs and goals, meaningful communication and inspiration, and a general policy of creating spaces for action involving varying degrees of freedom.

From an organization's point of view, it is legitimate and often necessary to initiate change top down, to define concrete goals, and to name the preferred procedures to achieve those goals. This is done by requiring employees to accept other-directed accommodation. The goals of a new culture of occupational safety and health (BMAS 2013; DGUV 2015) or an agile transformation are good examples of this. What matters is that employees engage in self-directed processes to concretize and live these abstract other-directed requirements or stimuli.

This scheme may be applied both to the competence acquisition of individuals and to that of organizations as self-directed actors. Binding legislation such as the aforementioned Noise and Vibration Safety and Health Regulation is an other-directed requirement implying different needs for change for different organizations.

4.2.3 ORGANIZATIONAL CONCEPTS

Although people acquire competences in a self-organized manner, they never do so in isolation. Organizational phenomena provide relevant contexts for competence acquisition. Initially, the organization serves as the focus point for the individual's thinking and experience (Schreyögg 2008); in its "biography", the organization develops cognitive and affective patterns and structures that structure all further development of knowledge and values. Organizational strategy is the first concretization of that focus point (see Mields and Birner, Chapter 3 in this book).

In an analogy to Knut Illeris's learning theory, which refers to humans, Henning Pätzold identifies organizational knowledge and the corresponding knowledge management as an organization's cognitive side and organizational culture as its affective side. Both dimensions are in a dynamic, oscillating relationship with each other. At the same time, the organization is constantly interacting with its environment (Pätzold 2017, 46).

From a systems theory point of view, these aspects are emergent phenomena. Emergence means that a system's higher levels cannot be fully explained from its individual components:

> An order or a quality shall be called *emergent* when it can no longer be explained by aggregating its parts or by adding up the qualities of those parts. The most important case with regard to what we are concerned with here is the emergence of social systems, especially organizations, from the interactions of persons. According to the criteria of emergence, organizations must hence have qualities that cannot be derived from persons alone.
>
> **(Willke 2004, 17, emphasis in the original,**
> **author's translation)**

Although these higher system levels are necessarily dependent on humans, they are characterized by autonomous laws, a specific form of self-organization. With regard to communication, Willke notes:

> This means that communication always involves humans; their participation is both necessary and indispensable. (…) Humans do not, however, control – let alone determine – communication. That is because much more is involved than just the possible intentions of humans as actors: the laws governing language and thought come into play, and most importantly the laws and logics of established communication patterns, which range from fleeting conventions to historically stable institutions (…). Through repetition and confirmation, certain forms may solidify into semiologies of signs and into semantics of forms, thereby achieving a certain degree of stability and a likelihood of being re-used.

(Willke 2014, 45–46, author's translation)

Organizational systems and an organization's employees do not reciprocally determine each other. But each system presents its own complexity, its own knowledge, and its own interpretations as stimuli to the other system. These stimuli serve as a matrix and an opportunity for individual as well as organizational competence acquisition, but their processing takes place in a self-organized manner based on the organism's own structure. What matters is that organizational and individual schemes and patterns can be linked to each other.

Phenomena extensively discussed in organizational theory, such as the distinction between an organization's official views ("espoused theory") and the views that actually guide its actions ("theories in use") (Argyris and Schön 1978; Sausele-Beyer 2011), or between "formal structures" and "actual work activities" (Meyer and Rowan 1977, 341), are evidence of the self-organized nature of an organization and its employees. At the same time, such informal structures serve as a resource for mutual understanding and structures that have proven to work in practice.

4.3   COMPETENCE MANAGEMENT

The following section starts off by presenting the classic top-down conception of competence management (4.3.1), followed by a discussion of why this conception is increasingly reaching its limits and why agile forms of competence management are necessary. It also presents the key dimensions of such agile competence management (4.3.2).

4.3.1   TOP-DOWN COMPETENCE MANAGEMENT

The competencies of an organization's employees are considered essential for the organization to accomplish its goals (Schreyögg and Eberl 2015, 13). Competence management is about systematically connecting an organization's goals with the competencies of its employees, taking account of both the organization's goals and the needs of employees (Bartscher et al. 2012; North et al. 2018). *Strategic competence management* refers to the identification of competencies that will be needed in the future, whereas *operative competence management* refers to the identification of the competencies that already exist within the organization, as well as the development of missing competencies (North et al. 2018, 16ff.).

Competence management has traditionally been conceived as a top-down process encompassing the following steps:

(1) Defining organizational strategy: This first step is usually the responsibility of senior management, although lower levels in the organization's hierarchy may be involved to varying degrees and in various forms. In most cases, defining strategy in that manner implies both an analysis of the organization's environment(s) and an analysis of the organization's resources, for instance, when strengthening corporate social responsibility or in the context of building a functioning in-house occupational safety and health organization.

(2) Identifying the required competencies: The required competencies may be identified based on the wording in the job descriptions. Here, organizations can draw on their own expertise, take guidance from elaborate scientific catalogs of competencies, or adopt best practice solutions. Competencies are broken down into facets of competence and specified into behavioral anchors. Healthy leadership competence, for example, is usually required in the job profile of any senior manager. The goal of "reducing employee absenteeism" builds on behaviors such as social support, recognition, and interest.

(3) Measuring the competencies available: Next, the competencies available within the organization are measured, followed by a comparison of targets and actual availability. For this kind of analysis, organizations can draw on a wide range of tried-and-tested diagnostic tools, such as the "Key Leadership Factors KLF" (Olofsson and Frintrup 2017, 478–499) or KODE/KODEX (Heyse 2017, 245–273).

(4) Capacity building: The organization must decide whether to buy or develop missing competencies. When it comes to competence development, one widely accepted insight that has emerged from the debates on competence is that limiting the teaching of competencies to classroom settings will not produce the desired outcomes (Staudt and Kriegesmann 2000). Rather, successful competence development benefits from aspects of self-directed learning in informal learning processes integrated into learners' work environment.

(5) Evaluation: The final step is an evaluation of the success of the competence development measures. The entire competency model is reviewed at regular intervals to ensure its appropriateness, combined with a review of organizational strategy.

In summary, a top-down conception of competence management fits an organization's need for control. The environment is complex but managing organizational goals and the related competencies in a systematic and precisely adjusted manner makes complexity appear more manageable and suggests clearly defined cause-and-effect relationships. The underlying management logic corresponds to a command-and-control mindset.

From the point of view of an organization's employees, a top-down approach to competence management provides helpful guidance in a complex VUCA world (Millar et al. 2018). Yet modeling the processes and conditions that govern the mental and physiological ways in which employees experience the contents of their work, especially in the context of occupational safety and health, is a complex task. The benefits of a top-down approach, such as guidance and possibly perceived competence at the subjective level, are countered by the risk of experiencing restrictions in autonomy. The higher the level of detail in the wording of competence requirements and management goals, the higher the risk that the development process produces other-directed qualifications rather than self-organized competences, failing to fully exploit the potential of autonomous action in open-ended contexts.

From the perspective of a culture of prevention, it is essential to distinguish between two approaches. Clearly defined and binding legislative acts and regulations connect very well with the top-down approach. If employees can be encouraged to perceive safety- and health-related policies not as other-directed restrictions of their autonomy but as measures that help them to (fully) achieve the organization's goals, this perspective may already become part of an organization's strategy and can be implemented in concrete competency models (see Schöbel, Chapter 5 in this book). By contrast, the concept of implementing a culture of occupational safety and health in a shared effort of all participating institutions and organizations is more complex and should build on the idea of (organizational and individual) self-organization. That process, however, takes place at the expense of precise management.

Finally, a VUCA world requires organizations to respond faster and faster to environmental changes. The top-down conception of competence management presented here is not sufficiently dynamic to produce such responses. A better way to meet this challenge is the concept of agility, which is discussed in the next section.

4.3.2 Agile Competence Management

Our world has always been complex. In recent decades, however, the degree of complexity has increased even more as a result of rising interconnectedness. And as soon as complex problems are addressed with complex solutions, feedback loops emerge that create more complexity. At the same time, complexity is associated with uncertainty and unpredictability. No computations could have predicted the smartphone, but it changed the rules of the game in a fundamental way. What is more, these complex and unpredictable problems call for ever faster processing.

The answer to this challenge given by organizations is agility. If reliable predictions of the future are impossible, organizations have to be in a position to offer flexible responses to environmental changes. The key focus in this context is the needs of an organization's customers. Agile organizations provide their services based on an ongoing conversation with their customers (see Mields and Birner, Chapter 3 in this book). This means that employees working in areas where the organization intersects with its environment are not only sources of highly relevant knowledge, it is also essential that they are capable of acting in a fast and flexible manner.

It is important to note in this context that there is no single concept of agility; numerous variants are found in theory and practice.

One crucial milestone in agile thinking was the 2001 "Manifesto for Agile Software Development" (Beck et al. 2001), in which the authors identified four major principles of agile software development. In agile software development, the focus is (1) more on individuals and interactions than on processes and tools, (2) more on working software than on comprehensive documentation, (3) more on customer collaboration than on contract negotiation, and (4) more on responding to change than on strict adherence to a plan (ibid.). The authors point out that while both sides of each principle are important, they value the new principles over the old ones.

Steve Denning (2016) understands agility as a combination of various management practices that represent a departure from traditional management models: agility as "a radical alternative to command-and-control-style management".

The most frequently applied agile method is "scrum". Scrum – developed by Jeff Sutherland and Ken Schwaber at the beginning of the 1990s – is a project management method involving an iterative process in which the project team realizes the most relevant subgoals in two-to four-week "sprints", then evaluates the outcomes and flexibly adjusts the next steps in ongoing conversations with the customer (Sauter et al. 2018, 35–37). Other prominent agile tools include the methods of "Kanban", which call for using a Kanban board to visualize and prioritize individual work packages (ibid.), or of "design thinking", which enables creative and innovative solutions for customer needs (ibid., 47–49).

The key point of departure is an "agile mindset", characterized by three basic principles: (1) the needs of the customer are the paramount concern of all employees of an organization, (2) decisions are made where the relevant competence is located (this point was addressed early on by Karl E. Weick and Kathleen M. Sutcliffe (2001)), and finally (3) it is crucial to adapt to change as quickly as possible.

This means it is certainly possible or even necessary that strategic decisions be made at the center of an organization, but their operational implementation is carried out by comparatively small, interdisciplinary, and self-organized teams working where the organization meets its environment. This requires a high degree of organization, but the focus shifts from top-down management towards the facilitation of self-organization. To make this happen, the organization needs to break down hierarchies and create degrees of freedom; likewise, it needs to foster an agile culture.

When it comes to competence management in a VUCA world, agility means that it is no longer possible to fully anticipate the competencies that will be needed in the future. The "future" may refer to conversations with a customer during a design thinking process, which may require the company to introduce a collaborative virtual reality (VR) or augmented reality (AR) application. Or it may refer to the unregulated, disruptive collapse of traditional workplace relationships brought on by digital platforms.

Introducing agility to competence management may involve three dimensions: participation, dynamics, and self-organization. These dimensions are not new, but they may mean substantial change depending on how strongly they factor into the process. Each dimension is discussed below.

(1) The *participation* dimension refers to the involvement of an organization's employees in the decision-making process, for instance, when it comes to introducing occupational safety and health policies (e.g., sustainability in the supply chain; see Waterman, Chapter 8 in this book). Its concrete manifestation is found somewhere between the two poles of no participation and regular and institutionalized participation. In a setting where competencies are identified and managed from the top down, relevant competencies are usually overlooked to various extents. Employees can become experts in

their competencies if they are involved in the identification and specification of comprehensive basic competences. This is not a new insight and
probably the least radical element of agile transformation. It does become
relevant, however, as soon as fundamental strategic decisions resulting in
competence requirements must be made at the center of the organization.
The effectiveness of such central decisions should not be overestimated,
however; neither should the effectiveness of decisions made at the borders
of the organization to their environment be underestimated.

(2) The *dynamics* dimension refers to the frequency and speed of change in the
sense of adjusting to environmental conditions. Its concrete manifestation is
found somewhere between the two poles of long-term planning ("five-year
plan") and real-time adjustments. If employees make relevant contributions
to required competences, and these contributions first have to be approved
by a multitude of stakeholders, such "bottlenecks" make any adjustment
highly time-consuming. Reducing hierarchical levels can help to make the
process more dynamic and may help employees experience a higher degree
of freedom. Likewise, dynamization may imply reducing the intervals for
evaluations and adjustments. Finally, dynamization means leaving some
resources free for unexpected changes. This does not mean that resources
remain unused, as such changes will occur on a regular basis, they merely
cannot be fully anticipated or planned.

(3) The *self-organization* dimension refers to the extent to which agile teams
are authorized to make decisions. Its manifestation is found somewhere
between the two poles of no authority and full decision-making authority,
including appropriate resources. Self-organized teams are the most effective form of dynamizing an organization and integrating the competence
of employees. However, it should be borne in mind that the development
of (some) competences may take a longer period of time, so that even
self-organized teams are not able to adapt constantly to changes in "real
time". The composition of expert teams in various subject areas helps to
ensure the internal integration and development of relevant competences
as quickly as possible. Self-organization and competence management
means enabling employees to identify and realize their needs for competence development in a self-organized manner. In this context, it is
important not to underestimate the effectiveness of social control when
it comes to ensuring the necessity of competence development measures.
This should not be interpreted too narrowly, however. Sometimes an organization's future success is ensured by competences whose necessity is
not immediately evident.

Agility creates an even stronger integration of learning and work than before (Sauter
et al. 2018). This implies a further increase in the core role of learning, especially
self-directed learning, in the work process. But self-organization in competence
development does not mean that employees are not allowed to, and/or do not want
to, ask for support. Here, human resource officers must serve as learning needs advi-
sors, which may mean shifting the focus from a management perspective towards

TABLE 4.1
Dimensions of Agile Competence Management

Dimension	Participation	Dynamics	Self-organization
Poles	No participation all the way to regular and institutionalized participation	Long-term planning ("five-year plan") all the way to real-time adjustments	No authority all the way to full decision-making authority including appropriate budgets

self-understanding as in-house service providers. Likewise, the organization's local leaders can become service providers facilitating this self-organization and nurturing the growth of their employees.

An agile organization must trust in the self-organized abilities of its employees and give these abilities sufficient room to unfold (Denning 2016). This may require understanding an organization's transformation towards agility as a developmental process, which may certainly be initiated and directed by senior management. Ultimately, the goal is to reduce or modify aspects of top-down management. Decision-makers may benefit from some professional coaching to become more aware of the core convictions and values that guide their own actions and to see how well they fit with an agile mindset. At the organizational level, it may be helpful to perform an assessment of the organization's culture. Individual and organizational values are mutually dependent, but they do not determine each other. Agile competence management always has to integrate aspects of organizational, team-related, and individual development (Sauter et al. 2018).

In summary, it can be said that from an organization's perspective, it seems necessary to identify, record, and develop required competencies in a reliable and systematic manner. In a dynamic and complex VUCA world, however, that approach is problematic. In an age of accelerated digitalization and in the face of rising so-called "artificial intelligence", it is no longer possible to predict the competencies that will be needed in the future. As a consequence, it becomes more important for organizations to redesign their work relationships based on partnership and to introduce suitable approaches to competence management.

For an organization's employees, agility means a higher degree of both autonomy and responsibility. The familiar guidance coming in the form of clear messages from management is no longer available and has to be replaced with shared values and visions and an agile culture. Complexity can only be reduced to a certain extent and must hence be tolerated and addressed. This complexity has consequences for organizational learning. Since the beginning of the competence debate, the importance of work-integrated learning has risen steadily. Work-integrated learning enables employees to directly observe, experience, and reflect on the consequences of their own actions. If everything is connected to everything else, however, the consequences of our actions often take effect at a later time or in a different place, making it difficult to identify linear connections. It remains to be seen how this will affect the perceived competence (Deci and Ryan 2000), the self-efficacy expectations

(Schwarzer and Jerusalem 2002), and ultimately the mental health of an organization's employees.

An organization's leaders are faced with the vital necessity of developing trust in their employees' capacities for self-organization. This is neither trivial, nor is it a linear process. Trust, self-organization, and success, or the opposite concepts of each, may be conceived as mutually reinforcing feedback loops. In terms of leadership, the focus shifts away from top-down management towards involvement and facilitation, with leaders serving as trainers of their staff.

4.4   CONCLUSION

Proponents of agility sometimes give the impression that agility can and must create continuous innovation (Denning 2016). Yet the overarching goal of agility is to address environmental requirements effectively and efficiently. To achieve this goal, it may be helpful to alternate periods of innovation and consolidation. Brian J. Robertson, who devised the concept of holacracy – a concept which decentralizes decision-making powers on the basis of a detailed system of rules and roles – describes a successful period in his agile company during which the primary focus in decision-making situations was on documentation and standardization rather than on innovation and transformation (Robertson 2015). In this context, it is important to remember that even professional athletes cannot perform an unlimited number of "sprints" in rapid succession. It helps to keep in mind that the dimensions of self-organization and dynamics are not necessarily tied to each other; they rather represent variable parameters for implementing agility in specific organizations and specific situations. This is equally true for agile competence management and for an agile organization overall.

From the perspective of occupational safety and health, it is essential to set binding standards, such as the German Dangerous Substances Order (GefStoffV), when it comes to fundamental aspects of safety and health and to obligate organizations to comply with these standards, especially in a context of self-organization and agile management. After all, it is up to an organization's employees to implement these standards on a daily basis, and a lack of statutory regulation involves high risks in terms of occupational safety and health. Too much regulation, however, may severely limit aspects of autonomy and self-organization, which may conflict with the goals of agile work. Finding a good balance here is anything but trivial.

Similar things may be said about the makeup of workplace relationships. If organizations think of their employees primarily as a means to achieve organizational ends, regulations with respect to working time and working time recording, for example, offer employees necessary protection. If organizations understand that agile working arrangements only succeed if management and employees work together as partners, standardized and binding policies ultimately limit the opportunities for self-organized and productive work for those involved.

Finally, complex competences, such as healthy leadership competence, are "dispositions for self-organization" (Erpenbeck 2013). They are developed in a self-organized manner and become manifest in autonomous actions in complex contexts. Binding standards should successively be replaced with opportunities for competence development, and they should be embedded in a culture of occupational safety and health that is based on partnership.

Right now, it is impossible to predict how work and society will evolve in the future and whether employees will be perceived as equal partners or as means to an end. The same is true concerning the issue of finding a balance between a top-down approach to designing safety- and health-related policies and an approach based on employee involvement and partnership. With respect to occupational safety and health, agility may ultimately mean to engage all relevant stakeholders in a conversation, to pay close attention to developments, and to act based on those observations. In that effort, it will be helpful to distinguish between areas in which binding standards are indispensable and areas in which cooperative self-organization is desirable. The more organizations become increasingly responsible actors with regard to occupational safety and health, the more top-down management styles can be supplemented or replaced with a culture of prevention. Ultimately, in this as in any other context, self-organization may mean that the autonomous actions of the partners involved may diverge from one's own ideas, but that the sum of these actions is more likely to achieve an organization's goals than consistent top-down management, which often does not accurately match the specific local situation.

REFERENCES

Argyris, C., and D. A. Schön. 1978. *Organizational Learning: A Theory of Action Perspective.* Reading, MA, et al.: Addison-Wesley.

Arnold, R. 2000. Qualifikation. In: *Wörterbuch Erwachsenenbildung*, eds. R. Arnold, S. Nolda, and E. Nuissl, 269. Bad Heilbrunn: Julius Klinkhard.

Arnold, R. 2010. Kompetenz. In: *Wörterbuch Erwachsenenbildung*, eds. R. Arnold, S. Nolda, and E. Nuissl, 172–173. Bad Heilbrunn: Julius Klinkhard.

Arnold, R., and C. Gómez Tutor. 2007. *Grundlinien einer Ermöglichungsdidaktik. Bildung ermöglichen – Vielfalt gestalten.* Augsburg: ZIEL Verlag.

Bartscher, T., J. Stöckl, and T. Träger. 2012. *Personalmanagement. Grundlagen, Handlungsfelder, Praxis.* Hallbergmoos: Pearson Deutschland.

Beck, K., M. Beedle, A.van Bennekum, A. Cockburn, W. Cunningham, M. Fowler, et al. 2001. Manifest für agile Softwareentwicklung. https://agilemanifesto.org/iso/en/m anifesto.html (accessed December 12, 2019).

Becker, M. 2013. *Personalentwicklung. Bildung, Förderung und Organisationsentwicklung in Theorie und Praxis.* 6th ed. Stuttgart: Schäffer-Poeschel.

BMAS – Federal Ministry of Labor and Social Affairs. 2013. *Empfehlungen für eine neue Kultur der Gesundheit im Unternehmen. Deutschlands Wettbewerbsvorteil.* https://ww w.bmas.de/SharedDocs/Downloads/DE/PDF-Publikationen/a860-gesundheit-im-unte rnehmen.pdf?__blob=publicationFile&v=2 (accessed December 12, 2019).

BMAS – Federal Ministry of Labor and Social Affairs. 2018. *Sicherheit und Gesundheit bei der Arbeit – Berichtsjahr 2017. Unfallverhütungsbericht Arbeit.* https://www.baua.de/ DE/Angebote/Publikationen/Berichte/Suga-2017.pdf?__blob=publicationFile&v=13 (accessed December 12, 2019).

Bollmann, U. 2018. Competences for a culture of prevention. Conditions for learning and change in SMEs. In: *Safety Management in Small and Medium Sized Enterprises (SMEs)*, eds. G. Boustras, and F. Guldenmund, 122–138. Boca Raton: Taylor & Francis.

Denning, S. 2016. The age of agile. *Strategy & Leadership* 44(4), 10–17.

DGUV – German Social Accident Insurance. 2015. *Strategic Concept for the Next Joint Prevention Campaign Implemented by the DGUV and Its Members.* Berlin: DGUV. https://www.dguv.de/medien/inhalt/praevention/kampagnen/praev_kampagnen/ausblick/fachkonzept_eng.pdf (accessed December 12, 2019).

Erpenbeck, J. 2013. Was "sind" Kompetenzen? In: *Bildung. Kompetenzen. Werte*, eds. W. G. Faix, J. Erpenbeck, and M. Auer, 297–353. Stuttgart: Steinbeis.

Erpenbeck, J., L. von Rosenstiel, S. Grote, and W. Sauter. 2017. *Handbuch Kompetenzmessung. Erkennen, verstehen und bewerten von Kompetenzen in der betrieblichen, pädagogischen und psychologischen Praxis.* 3rd ed. Stuttgart: Schäffer-Poeschel.

Hahnzog, S. 2015. *Gesunde Führung. Impulse für den Mittelstand.* Wiesbaden: Gabler.

Heuser, M.-L. 2016. Autopoiese und Synergetik. In: *Synergetik. Kultur- und Wissensgeschichte einer Denkfigur*, eds. T. Petzer, and S. Steiner, 149–163. Paderborn: Wilhelm Fink.

Heyse, V. 2017. KODE® und KODE®X – Kompetenzen erkennen, um Kompetenzen zu entwickeln und zu bestärken. In: *Handbuch Kompetenzmessung. Erkennen, Verstehen und bewerten von Kompetenzen in der betrieblichen, pädagogischen und psychologischen Praxis*, eds. J. Erpenbeck, L. von Rosenstiel, S. Grote, and W. Sauter, 3rd ed. 245–273. Stuttgart: Schäffer-Poeschel.

Illeris, K. 2006. Das "Lerndreieck". Rahmenkonzept für ein übergreifendes Verständnis vom menschlichen Lernen. In: *Vom Lernen zum Lehren. Lern- und Lehrforschung für die Weiterbildung*, ed. E. Nuissl, 29–41. Bielefeld: Bertelsmann.

Illeris, K. 2010. *Lernen verstehen. Bedingungen erfolgreichen Lernens.* Bad Heilbrunn: Klinkhardt.

Klieme, E., K. Maag-Merki, and J. Hartig. 2007. Kompetenzbegriff und Bedeutung von Kompetenzen im Bildungswesen. In: *Möglichkeiten und Voraussetzungen technologiebasierter Kompetenzdiagnostik. Eine Expertise im Auftrag des Bundesministeriums für Bildung und Forschung*, eds. J. Hartig, and E. Klieme. Bonn: BMBF.

Knowles, M. S. 1975. *Self-Directed Learning. A Guide for Learners and Teachers.* New York: Association Press.

Krohn, W., and G. Küppers. 1990. Preface. In: *Selbstorganisation. Aspekte einer wissenschaftlichen Revolution*, eds. W. Krohn, and G. Küppers, 1–17. Braunschweig / Wiesbaden: Vieweg & Sohn.

Maturana, H. R. 1985. *Erkennen. Die Organisation und Verkörperung von Wirklichkeit.* 2nd ed. Braunschweig / Wiesbaden: Vieweg.

Meyer, J. W., and B. Rowan. 1977. Institutionalized organizations: Formal structure as myth and ceremony. *American Journal of Sociology* 83(2), 340–363.

Millar, C. C. J. M., O. Groth, and J. F. Mahon. 2018. Management innovation in a VUCA world: Challenges ad recommendations. *California Management Review* 61(1), 5–14.

North, K., K. Reinhardt, and B. Sieber-Suter. 2018. *Kompetenzmanagement in der Praxis. Mitarbeiterkompetenzen systematisch identifizieren, nutzen und entwickeln. Mit vielen Praxisbeispielen.* 3rd ed. Wiesbaden: Springer Fachmedien.

Olofsson, A., and A. Frintrup. 2017. Key leadership factors KLF. In: *Handbuch Kompetenzmessung*, eds. J. Erpenbeck, L. von Rosenstiel, S. Grote, and W. Sauter, 478–499. Stuttgart: Schäffer-Poeschel.

Pätzold, H. 2011. *Learning and Teaching in Adult Education: Contemporary Theories.* Opladen et al.: Budrich.

Pätzold, H. 2017. Das organisationale Lerndreieck. Eine lerntheoretische Perspektive auf organisationales Lernen. *Zeitschrift für Weiterbildungsforschung* 40(1), 41–52.

Pätzold, H., and F. Brendebach. 2018. Erwachsene in der Berufsbildung. In: *Handbuch Berufsbildung*, eds. R. Arnold, A. Lipsmeier, and M. Rohs, 109–120. Wiesbaden: Springer VS.

Robertson, B. J. 2015. *Holacracy.* New York: Henry Holt.

Sausele-Bayer, I. 2011. *Personalentwicklung als pädagogische Praxis*. Wiesbaden: VS-Verlag
    für Sozialwissenschaften.
Sauter, R., W. Sauter, and R. Wolfig. 2018. *Agile Werte- und Kompetenzentwicklung. Wege in
    eine neue Arbeitswelt*. Berlin, Heidelberg: Springer.
Schreyögg, G. 2008. *Organisation. Grundlagen moderner Organisationsgestaltung. Mit
    Fallstudien*. 5th ed. Wiesbaden: Gabler.
Schreyögg, G., and M. Eberl. 2015. *Organisationale Kompetenzen. Grundlagen – Modelle –
    Fallbeispiele*. Stuttgart: Kohlhammer.
Schwarzer, R., and M. Jerusalem. 2002. Das Konzept der Selbstwirksamkeit. *Zeitschrift für
    pädagogik* 44, 28–53.
Staudt, E., and B. Kriegesmann. 2000. Trotz Weiterbildung inkompetent. In: *Jahrbuch
    Personalentwicklung und Weiterbildung 2000/2001*, eds. K. Schwuchow, and J.
    Gutmann, 39–44. Neuwied, Kriftel: Luchterhand.
Weick, K. E., and K. M. Sutcliffe. 2001. *Managing the Unexpected: Assuring High
    Performance in an Age of Complexity*. San Francisco: Jossey-Bass.
WHO – World Health Organization. 1986. *Ottawa Charter for Health Promotion*.
Willke, H. 2004. *Einführung in das systemische Wissensmanagement*. 4th ed. Heidelberg:
    Carl-Auer-Systeme Verlag.
Willke, H. 2014. *Regieren. Politische Steuerung komplexer Gesellschaften*. Wiesbaden:
    Springer Fachmedien.

5 Managing Competencies of Safety Leaders
Some Promises and Shortcomings

Markus Schöbel

CONTENTS

5.1 INTRODUCTION

Working safely in risky workplaces does not just mean not having accidents. Working safely is also about what an organization and its members do to prevent accidents. This means that operating technical systems safely is not merely about avoiding errors. It is also about being sensitive to early signs of failure, learning from previous accidents, and anticipating what might happen in the future (Rochlin 1999; Weick and Suttcliffe 2007). One important influencing variable initiating and supporting those safety-directed activities is leadership behavior. By definition, it refers to social interactions in a hierarchical relationship and describes interpersonal influence processes between a leader and co-worker(s). Past research has shown that leadership importantly drives organizational cultures that have to perform safely and reliably (Guldenmund 2007). Moreover, leadership influences on safety are regularly recognized in major accident investigations (Baker 2007; Fruhen et al. 2014), and empirical studies have revealed that leadership performance has a crucial impact on employees' compliance with safety rules and voluntarily participation in safety initiatives (Christian et al. 2009; Martínez-Córcoles et al. 2013; Donovan et al. 2016).

This chapter draws on these previous findings and tries to explore the potential of identifying and developing leadership competencies that contribute to safer workplaces. This is done with a strong focus on "competency modeling", a powerful technique that is already implemented in many of today's organizations. It follows the idea of aligning the goals, visions, and strategies of an organization with the individual behavior of its members, and thereby providing benchmarks for monitoring and developing members' behavior in a desired direction. Although there is a vivid debate and much skepticism about the benefits of competency modeling in organizations (Conger and Ready 2004; Hollenbeck et al. 2006; Campion et al. 2011), one has to take into account that due to its extensive use in the field, it may provide an important and accepted way of systemically improving safety leadership behavior and, therefore, the safety culture of an organization. Given that empirical evidence on safety leadership competencies – or on competency modeling in hazardous workplaces per se – is very scarce (as will be shown later on), this chapter breaks mainly new theoretical ground by exploring the potential of this technique for promoting safety. This is done by first introducing and defining the concept of competency modeling and associated constructs in general. In a next step, an exemplary framework of generic safety leadership competencies is developed. This is specified according to findings from a literature search on empirical evidence identifying facets of safety leadership behavior having a positive impact on safety outcomes. After that, potential benefits and shortcomings of managing safety leadership competencies in the field are discussed. Finally, the challenges and future research needs of competency modeling in hazardous workplaces are outlined.

5.2 DEFINING (LEADERSHIP) COMPETENCIES AND COMPETENCY MODELING

Competency modeling is a common practice in organizations today. Creating fit and alignment between individual competencies and organizational strategies and objectives is assumed to play an important role for a sustained competitive advantage (e.g., Ulrich and Smallwood 2004). Specifically, competency modeling allows organizations to translate their business strategies into performance requirements for their employees (Stevens 2013). By providing a *common language* in the form of desirable sets of individual competencies, it serves as a platform for a variety of human resource (HR) activities, by informing, for instance, selection, training and development, promotions, job assignment, and compensation (Rodriguez et al. 2002; Lievens et al. 2004).

In general, a competency model (i.e., a set of competencies) describes how individuals excel in specific job positions and responsibilities (Boyatzis 1982). Campbell and Wiernik (2015) identified three ways of conceptualizing competencies: a competency could refer to performance itself, to a direct determinant of performance (e.g., negotiating skill), or to a more distal indirect determinant of performance (e.g., openness to experience). Thus, behavioral outcomes are attributed to traits, motives, skills, or knowledge, which are needed for effective performance in the jobs, and/or, on the other hand, to performance itself (Stevens 2013). Whereas competency models differ enormously in their real-life conceptualizations (e.g., mainly presenting a mixture of behavioral and psychological facets), prominent scholars in the academic field favor the assumption that competencies are best classified as part of the performance space (Bartram 2005; Lievens et al. 2010) or as "behavioral themes" that are considered to be critical success factors and strategic performance drivers (Sanchez and Levine 2009; 2012).

One main psychological research focus lies on investigating transferable generic competencies that are required for most jobs or particular occupations or job roles. Kurz and Bartram (2002) developed a prominent competency framework with eight competency domains (also known as *the great eight*) depicting a behavioral collection that is necessary for the delivery of desired outcomes in organizations. Competencies are meant to reside in the behavioral act that leads to good performance outcomes implying that competencies can be learned and developed. The eight empirically derived competency domains for general work performance provided by Kurz and Bartram (2002) are as follows: Leading/Deciding; Supporting/Cooperating; Interacting/Presenting; Analyzing/Interpreting; Creating/Conceptualizing; Organizing/Executing; Adapting/Coping; Enterprising/Performing. These eight competencies are specified as higher-order factors representing 112 individual scales defining negative and positive behavioral indicators (Bartram 2005). With regard to leadership competencies, Kurz and Bartram's framework identifies the domain of *leading and deciding* as one of the "great eight" competencies, defined as "takes control and exercises leadership, initiates action, gives direction, and takes responsibility" (Bartram 2005, 1187). The competency itself contains several behavioral aspects of *providing leadership and supervision* (e.g., providing direction and coordinating action, supervising and monitoring behavior or coaching) and deciding and initiating action (e.g., making decisions, taking responsibility, or acting with confidence). However, understanding leadership as an interpersonal influence process (Yukl 2013), this differentiation does not allow disentangling leadership behavior from general management performance. *Deciding and initiating action* involves direct interpersonal influences as well, and these are not necessarily bound to hierarchical relationships.

An alternative competency framework was developed by Campbell (2012; Campbell and Wiernik 2015). It provides an integrative synthesis of eight general content dimensions of performance in a work role meant to integrate previous work on individual performance modeling, team member performance, and leadership and management performance. Each dimension consists of several subfactors highlighting the description of competencies as concretely as possible. Contrary to the framework of Kurz and Bartram (2002), Campbell and Wiernik (2015) explicitly differentiate between leadership performance and management performance in hierarchical relationships. In their framework, leadership refers to the interpersonal influence process, whereas management performance deals with actions such as generating, preserving, and allocating the organization's resources to best achieve its goals (Campbell and Wiernik 2015, 54). Specifically, the competency domain of *hierarchical leadership performance* contains six behavioral facets referring to what leaders do: (1) consideration, support, person-centeredness; (2) initiating structure, guiding, directing; (3) goal emphasis; (4) empowerment, facilitation; (5) training, coaching; and (6) serving as a model. Although emphasis on each subfactor and associated actions may vary at different organizational levels and settings, Campbell and Wiernik (2015) observed a striking convergence of research literature on leadership models ranging from the classical Ohio State and Michigan studies to current leadership theories such as transformational leadership or team leadership (ibid.).

To the best of my knowledge, research findings on competencies for safety leadership are rarely published in the academic literature. Instead, a reasonable range of competency models are presented, which either were developed with regard to specific work contexts and job roles (e.g., nursing leadership competency model by Sherman et al. 2007) or describe more general competencies of safety professionals without focusing on leadership positions (Chang et al. 2012). Therefore, the next section describes the attempt to develop an exemplary outline of a generic competency model for "Safety Leadership", based on research evidence stemming from review papers (e.g., meta-analyses or theoretical reviews) and empirical field studies on safety leadership. This model was developed beyond a single organization or industrial domain and provides a generic structure identifying common safety leadership competencies across various organizations and levels of leadership positions. Importantly, it does not take into account competencies, (1) which are important for all employees to possess (so-called *core competencies*, Hamel and Prahalad 1994), (2) which describe job-specific skills required to perform a particular job profession (*functional competencies*, Ozcelik and Ferman 2006), and (3) which focus on management performance or, in other words, the generation, preservation, and allocation of resources. Thus, the developed model has a relatively narrow focus and should inspire further refinement with regard to a given industrial domain or organization.

5.3 DEVELOPING A GENERIC SAFETY LEADERSHIP COMPETENCY MODEL

As a starting point for developing a generic competency model of safety leadership, Campbell's (2012) conceptualization of hierarchical leadership performance in terms of six subfactors was used. In order to formulate and assign specific safety leadership competencies to the six subfactors, a literature search was conducted focusing on empirical and theoretical evidence on leadership models and competencies warranting safe workplaces. Adopting Bartram's (2005) definition, competencies were understood as something that people actually do and that can be observed. Thus, the aim was to extract the behavior-oriented competencies of safety leaders from the identified research literature.

In a first step, "safety leadership competency" was used as keyword in the Web of Science (Clarivate Analytics 2018); however, this yielded no results. Then, other keywords (such as "safety leadership" and "leadership for safety") were chosen to search for safety leadership competencies and 142 articles were found. These were complemented by five additional articles that were cited by the initially found articles and matched the scope of the literature search. Confining the search to articles that were published in peer-reviewed journals yielded a total of 109 articles. From these articles, 52 were excluded because they focused on systemic safety improvement strategies without specifying concrete competencies or practices associated with a leader's behavior. Finally, a total of 57 relevant papers were included; specifically, 51 articles and six review papers investigating the impact of specific leadership models on safety performance were selected. These models investigated are rooted in general leadership research and were applied over a broad range of industrial domains where safety is at stake (e.g., nuclear power plants, military organizations, or medical settings). Within these 57 articles, the following leadership models are considered (note that some articles investigate more than one model): transformational-transactional leadership (e.g., Clarke 2013) ($n = 25$ studies); leader-member exchange (e.g., Hofman and Morgeson 1999) ($n = 7$ studies); safety leadership model (e.g., Wu 2005) ($n = 14$ studies); leader walk-round (e.g., Sexton et al. 2014) ($n = 6$ studies); empowering leadership (e.g., Martínez-Córcoles et al. 2012) ($n = 8$ studies); high reliability leadership (Martínez-Córcoles 2018) ($n = 1$ study); supervisor enforcement (e.g., Petitta et al. 2017) ($n = 2$ studies); authentic leadership (e.g., Eid et al. 2012) ($n = 3$ studies); and "mixed" leadership models (e.g., Griffin and Hu 2013) ($n = 6$ studies).

In a next step, two raters independently assigned leadership competencies from the models (extracted from the subdimensions of identified models and corresponding measurement scales) to the six subfactors of the Campbell framework. After that, the assignments were discussed in a group meeting, where differences in assigning the competencies to the six subfactors were collaboratively resolved. Competencies were formulated by focusing on interpersonal safety-relevant dynamics between a leader and co-workers (i.e., "how a leader influences co-workers to work safely"). It was aimed at formulating the competencies as concretely and specifically as possible and to minimize their number. In a last step, the developed competency framework was discussed in a group of 15 safety field experts, who refined the formulated competencies with regard to their comprehensibility and checked the content validity of the developed competency model. In Table 5.1 an outline of the model is presented describing each of the six competency domains according to Campbell (2012) with its assigned competencies.

The first domain refers to a leader's consideration and support for co-workers. It describes competencies demonstrating concern (by providing recognition and encouragement with regard to safe work performance) and showing an understanding to build trust and sound relationships (e.g., by helping and giving constructive feedback). The competencies were mainly derived from the transformational and empowering leadership model. More specifically, they describe a leader's individualized consideration of team members and his/her behavior in acknowledging team efforts, demonstrating a general regard for team members' wellbeing, and keeping track of what is going on in the team. Moreover, these competencies also involve an open sharing of a leader's own thoughts and beliefs about safety in the workplace.

The second domain refers to a leader's guiding and directing behavior. The designated competencies describe a leader's behavior aimed at organizing and coordinating co-workers' safe task fulfillment, which involves clarifying work roles and providing the necessary resources in terms of technical support and knowledge. The designated competencies mainly stem from transactional leadership practices; namely, clarifying expectations and rewards in exchange for meeting expectations (i.e., contingent reward) and monitoring behavior and taking corrective action prior to the occurrence of major safety problems (i.e., management by exception [active]).

TABLE 5.1

Safety Leadership Domains and Assigned (Performance-Focused) Competencies

Competency domains (according to Campbell 2012)	(Performance-focused) competencies of safety leadership
Consideration and support	• Encouraging co-workers to work safely by helping them to recognize the necessity of safety in their daily work performance • Providing feedback on co-workers' task performance, while recognizing their contributions and efforts to work safely • Showing concern by taking time to discuss co-workers' troubles and doubts regarding safe work performance
Guiding and directing	• Motivating co-workers in safety performance by rationalizing rule compliance and organizational goals and objectives • Pro-actively identifying hazard potentials by monitoring work performance and rule compliance (and in case of deviations, investigating non-compliance and root causes of incidents) • Clarifying expectations by clear communication of co-workers' roles and safety obligations
Goal emphasis	• Communicating the priority of safety over production goals by knowing what safe work performance looks like and by which factors it can be undermined • Communicating a safety vision in a language that resonates with the co-worker • Building commitment to (safe) work performance by instilling pride
Empowerment and facilitation	• Encouraging participation by communicating own beliefs and thoughts and appreciating co-workers' behavior to speak-up • Involving co-workers by actively listening to co-workers' opinions and beliefs about safety before reaching conclusions • Delegating responsibility to co-workers with regard to performance outcomes
Training and coaching	• Promoting rationality and problem solving by challenging subordinates with elaborate safety paradigms or new ideas aimed at a rethink of their own performance • Evaluating and meeting performance needs of co-workers • Actively coaching co-workers to foster safe collaboration and a team identity
Serving as a model	• Seeking feedback from co-workers with regard to own leadership performance • Admitting own mistakes and weaknesses regarding safety • Showing visible and consistent safety commitment by demonstrating a safety-driven focus for delivering results

The third domain describes a leader's goal emphasis. It involves a leaders' behavior in encouraging enthusiasm and commitment (for the groups' and organizational safety goals and visions) and motivating others to high expectations to build a higher aspiration for safety. Here, one facet of the transformational leadership model (i.e., inspirational motivation) was crucial for the assignment of these competencies.

The fourth domain refers to empowerment and facilitation. The assigned competencies describe a leader's behavior of encouraging team members' safety participation behavior by fostering the expression of ideas and opinions and delegating authority and responsibility. The competencies were mainly derived from the subdimensions "participative decision-making" of the empowering leadership model and "balanced processing" of the authentic leadership model. They describe how a leader uses team members' information in making decisions, solicits opposing viewpoints, and fair-mindedly considers those viewpoints.

The fifth domain relates to competencies with regard to training and coaching. The behavior in question involves coaching and instructing co-workers regarding the joint safe accomplishment of job tasks by providing coherence and a team identity. The competencies involve two subdimensions of transformational leadership (i.e., individualized consideration and intellectual stimulation) as well as the behavioral dimension coaching of the empowering leadership model.

The sixth domain describes competencies with regard to serving as a model. This means that a leader models appropriate behavior by exhibiting principled and ethical behavior. This also involves leaders building credibility by developing safety leadership from their own personality and values. The competencies stem from the transformational dimension of idealized influence as well as from the authentic and empowering leadership model.

This exemplary competency model was developed from research findings on safety leadership derived from a broad range of industrial domains. Bringing to bear its potential for improving an organization's safety culture, the model has to be tailored to organizational performance coping with specific hazards and to be integrated into organizational management systems (i.e., with regard to monitoring and developing leaders' competencies). However, both steps (i.e., mapping and managing competencies) are highly vulnerable to factors that may undermine the model's potential for improving safety leadership behavior. Thus, in the following sections some potential benefits and shortcomings of mapping and managing safety leadership competencies in applied contexts are outlined.

5.4 MAPPING AND MANAGING SAFETY LEADERSHIP COMPETENCIES IN ORGANIZATIONAL SETTINGS

Although safety leadership competencies may play only a minor role in organizational management systems compared with the management of core and functional competencies, they provide specific challenges when aligned to them. The following section outlines some benefits and shortcomings that become crucial when a safety leadership competency model is embedded in management systems.

Taking a macro perspective on managing competencies, the cycle of continuously enhancing and developing competencies can be roughly described according to four phases: mapping, diagnosing, developing, and monitoring (Draganidis and Mentzas 2006). *Mapping* provides the organization with an overview of all the competencies necessary to fulfill its goals and objectives. *Diagnosing* defines the gap between the number and level of competencies that leaders possess, in comparison with the number and level of competencies required by the organization according to their leadership position. *Developing* deals with scheduling of activities aimed at increasing the number and proficiency level of competencies that the leaders should have. *Monitoring* means that results achieved by the competency development phase are continuously examined.

Improving safety leadership competencies by integrating them within this management cycle provides several benefits for organizations. According to Conger and Ready (2004), leadership competency models offer in general three benefits: *clarity, consistency*, and *connectivity* (the authors also identified three more "Cs" that present limits to leadership competency models by being *complicated, conceptual*, and built around *current* realities). First, competency modeling helps organizations to develop *clear expectations* about the types of behaviors, mindsets, and values that are important and valued to those in leadership roles. They provide a clear picture of an ideal safety leader. As a consequence, every leader in the organization knows what to do to be effective and is informed about poor and good leadership performance. Because competencies are embedded in feedback tools, Conger and Ready (2004) state that managers can more easily ascertain where their strengths and development needs lie (simply by scanning a set of numerical scales resulting from feedback surveys). The second benefit is that competency modeling provides *consistency*. By establishing a single model for organizational leaders, this model provides a common framework and language for communicating and implementing an organization's leadership development plans. Thus, competency models allow structuring HR activities in a consistent manner and, therefore, streamlining the improvement of leadership behavior within the whole organization in the desired direction. A third related advantage is *connectivity* to other management processes. Leadership competencies are assumed to become a driving force in performance management and feedback processes, high-potential identification, succession management, and reward schemes (Conger and Ready 2004, 43).

However, these benefits face severe shortcomings that might undermine improving safety leadership behavior due to managing competencies. In the following, these shortcomings are grouped around two (clearly interwoven) questions: (1) how to map and align (ideal) leader competencies onto the organization's goals and objectives, and (2) how to diagnose, monitor, and develop these competencies.

5.5   POTENTIAL SHORTCOMINGS OF MAPPING LEADERS' COMPETENCIES

Implementing a competency model into organizational managements systems implies that an organization has a clear picture of an ideal leader. Consequently, the corresponding competency models describe an *universal best-in-class leader capable of functioning in all situations* (Conger and Ready 2004, 44), and thereby assuming uniformity of responsibilities and corresponding requirements for all leaders in and across organizations. For instance, the safety leadership framework developed here clearly makes such a "best-in-class leader" impression by describing ideal behavioral facets of leader behaviors. Even when it was not intended to provide a ready-to-implement safety leadership competency model, it became obvious that the competencies described can hardly function in all situations and maybe cannot be shown by one and the same leader. Hollenbeck and McCall (2006) have argued against this *best-in-class leader* view, which they see as highly problematic:

> Effective leaders are not the sum of a set of competencies, however long or broad the list. Leaders, like the rest of us, are particular mixtures of pluses and minuses, the effectiveness of which changes over time and with the circumstances. …, that effectiveness depends on how various combinations of strengths are used, and that different strengths and weaknesses come in and out of focus at different times in different jobs. What matters is not a person's sum score on a set of competencies, but how well a person uses whatever talents he or she has to get the job done. Leaders do not come in neat, additive packages.

(Hollenbeck and McCall 2006, 399f.)

From this it becomes evident that developing competency models by simply adding competencies will not lead to accurate descriptions of effective leaders. Research on safety leadership also shows that some competencies might be more important than others depending on the hierarchical position of a leader, situational contingencies, and temporal factors. Flin and Yule (2004) argued that roles and responsibilities of leaders differ according to their position in the organizational hierarchy. They differentiated between three levels of leadership (i.e., supervisor positions, middle and top management positions) and assigned different requirements to each level. For instance, a key competency for top managers lies in the domain of goal emphasis, which requires the articulation of a vision of future safety performance and/or the symbolical demonstration of personal commitment to safety. On the other hand, monitoring co-workers' safety performance with regard to the safe fulfillment of work tasks lies primarily in the area of responsibility of a direct supervisor or first-line manager (i.e., the competency domain of guiding and directing); implying that the influence of each competency on warranting safe workplaces differs according to the hierarchical position of a leader.

The importance of competencies for safety leadership also varies with regard to situational contingencies. The High Reliability Organization (HRO) leadership model of Martínez-Córcoles (2017) differentiates between leading expansion, i.e., leadership behavior, which seeks to expand followers' cognitive categories by increasing their alertness and creating conditions where they can freely report any failure or mistake detected (Martínez-Córcoles 2017, 242), and leading reaction, i.e., leadership behavior, which seeks to respond to immediate demands as quickly and effectively as possible providing fast situation-focused solutions (ibid.). This differentiation pays tribute to the different operating conditions under which safety leadership in organizations takes place, namely stable or predictable conditions versus unstable or unpredictable conditions (Baran and Scott 2010). Depending on those contingencies some competencies seem to be more important than others, or, in a more extreme case, some competencies may also have counterproductive effects with regard to leadership effectiveness. For instance, the competency "admitting own mistakes and weaknesses regarding safety" is a somewhat useful behavior in stable situations, helping leaders and co-workers to reflect upon their own behavior and its impact on safety. However, in unstable situations, this behavior seems to be counterproductive in that it may signal uncertainty in a leader whose main goal should be to safely navigate a team through that (unstable) situation.

A further shortcoming refers to the temporal dimension in competency modeling. Conger and Ready (2004) have argued that competency models mainly focus on *current* leadership behaviors. These models tend to stabilize themselves in organizational systems due to extensive investments, which are required to establish performance and feedback systems, and involve the time it takes to educate managers in these new models (Conger and Ready 2004, 46). However, the competencies that helped current leaders to be successful may not be appropriate for future leaders of the next generation. For instance, according to first outlines of the so-called industry 4.0 era, manufacturing processes will be increasingly converged with technological developments such as sensors, cyber-physical systems, the Internet of things, and smart devices (Prifti et al. 2017; see Mields and Birner, Chapter 3 in this book). With the introduction of these new technologies, organizational risk management is massively challenged by the need to identify new risk factors and the necessity to have safety experts available, who will be, on the other hand, less and less present on the shop floor (Badri et al. 2018). This could mean that a rigorous management of information will become a key success factor in managing system safety in the industry 4.0 era. Consequently, leadership behavior in terms of empowering and facilitating information flows will be one of the most important competencies during those transition processes.

In sum, the impact of a leader's single competency on promoting safety differs with regard to the hierarchical positions, situational demands, and temporal demands. When competency modeling should inform leaders and organizations on effective safety leadership, it has to take those factors into account. One way to overcome these shortcomings is a combination of traditional job analysis and competency modeling methods, allowing for a more robust approach to competency modeling (Campion et al. 2011). Competency modeling is less rigorous than job analysis in terms of data collection, level of detail, assessment of reliability, and documentation of the research process (Schippmann et al. 2000). And indeed, the relatively scarce empirical competency modeling research has shown that job analysis methods enrich competency modeling with regard to methodological concerns such as the accuracy, inter-rater agreement, and discriminant validity of competency ratings per se (Lievens et al. 2010; Sanchez and Levine 2009). For instance, a study of Lievens, Sanchez, and DeCorte (2004) reveals that the quality of inferences drawn in competency modeling is enhanced by two procedural "task analysis" factors: the provision of task information and the use of a variety of job experts representing different views or "takes" on a job (e.g., supervisors, or internal customers).

Whereas a job analysis is more an inductive method (i.e., the analysis starts with job tasks and the requirements for the job to arrive at conclusions about what is important to the job in terms of KSAOs [Knowledge; Skills; Abilities; Other characteristics]), competency modeling is more deductive (starting with the strategic goals and outcomes of an organization and backing into individuals' tasks and behavior). According to Campion et al. (2011), combining methods from both camps will result in (1) using data collection methods such as observations, SME interviews, and focus groups to identify potential competency information; (2) providing clear construct definitions in the competencies and linkages to theory and literature; and (3) using survey methodology to empirically identify the critical competencies, which differentiate between job grades, situational demands, and temporal demands (Lucia and Lepsinger 1999; Rodriguez et al. 2002; Campion et al. 2011).

One potential drawback in applying the results of these methods to competency models (e.g., in the sense of a model with "rated" and "weighted" competencies) could be that such models simply run the risk of being too complicated. Conger and Ready (2004) stress this problem, stating that some competency frameworks specify up to 30 desired leadership dimensions. From a practical standpoint, this makes it rather difficult to trace and foster the development of competencies in organizational leaders, which will be discussed in the next section.

5.6 POTENTIAL SHORTCOMINGS OF MONITORING AND DEVELOPING LEADERS' COMPETENCIES

Another group of shortcomings refers to processes of monitoring and developing safety leadership competencies. As mentioned above, managing competencies involves the identification of gaps (between actual and desired competencies) and keeping track of the development of competencies based on these gaps. Modern human resource management provides a broad range of methods through which competencies can be assessed such as 360-degree feedback, assessment or development centers, or surveys. These methods are meant to allow for evaluating developmental interventions by, for instance, assessing competencies before and after interventions. However, it may become a difficult endeavor when it comes to the assessment of safety leadership competencies.

First, there is the question of how to evaluate proficiency levels of leaders. On the one hand, it makes total sense to describe competencies as concretely as possible; that is, to describe competencies as behavioral components, which makes them observable and somewhat countable. On the other hand, it is hardly possible to define objective threshold levels for behavior or, to make it simple, a behavior that can solely be judged if it is shown or not. Thus, developmental programs, which are only built on the occurrence of behavior, seem to be too simple and not straightforward. In addition, they do not take into account that a leader may compensate for a lack of some competencies (e.g., by drawing effectively on the knowledge of others) or change the job so that the competencies are no longer required for effective performance (e.g., due to crafting given task demands; Lievens et al. 2010).

Another problem refers to the evaluation of leadership behavior with regard to its effects on safety. In general, safety is understood as a (dynamic) non-event (Weick 1987). Thus, the enlisted safety leadership competencies, if completely shown by leaders, should contribute to establishing normal safe system states. The problem is that demonstrating such safety competencies usually does not attract a lot of attention and recognition in organizations. Moreover, they are somewhat expected and seen as more or less normal compared to competencies associated with the attainment of production or economic goals (Reason 1990). This also directs the evaluation of leadership competencies to the occurrence of accident and incidents (e.g., according to accident statistics on team levels), which might also imply problems. This inevitable link (i.e., the leadership behavior-accident statistics-link) neglects, for instance, system approaches to organizational safety. Important contributing factors (e.g., resource allocation, regulatory styles) embedded in safety settings are ignored. Evaluating leadership performance according to accident statistics can even lead to counterproductive developments; for instance, by leading teams to "manage the indicator" (e.g., by raising reporting thresholds) or hiding safety-relevant information. In addition, leadership behavior always involves an interaction, and it should be therefore understood as a reciprocal process. Thus, even when a leader shows desired competencies, their positive impact on safety also depends on the reactions of the co-workers and their willingness to follow their leader. Relatedly, Hollenbeck and McCall (2004) state that inter-individual combinations of competencies matter and not just the individual competencies. They give the example of autocratic leaders, who can be quite successful if they work together with capable people and listen to them; on the other side, consensus leaders can be quite unsuccessful if they want to empower ineffective or self-serving people (Hollenbeck and McCall 2004, 407).

All in all, monitoring and developing safety leadership competencies is prone to the difficulty in reliably monitoring and therefore tracing leaders' competencies and their impact on safety. Consequently, Campion et al (2011) suggest that promoting and developing desired leadership competencies should be embedded in a broader organizational development intervention instead of solely relying on data collection and customizing efforts within management systems. They locate organizational development techniques at the core of competency modeling (compared with job analysis where it is usually a peripheral activity), which means that a widespread involvement of organizational members in developing a competency model is necessary to get people to use and accept the model. Based on practical experiences with competency modeling rather than on empirical evidence, the authors argue that organizational development is crucial at all stages of a competency-modeling project and its associated principles should be applied in conducting such a project. For instance, in the planning phase it is necessary to involve senior management. Thus, they will be committed to the project objectives and promoting it. Next, collecting data for diagnosing, developing, and evaluating competencies should rely on using organizational development techniques like structured brainstorming, Delphi, and nominal groups as well as using action research methods (survey feedback and action planning) to validate the model's domains. Finally, implementing and institutionalizing competency models should integrate the classical Organizational Development (OD) concepts of "unfreezing" and "refreezing" to communicate the importance of considering resistance to change and other related project challenges (Campion et al. 2011, 251). In sum, the application of organizational development principles and tools in developing and implementing competency modeling holds a lot of promise because these techniques also importantly focus on both employee satisfaction and organizational effectiveness (Campion et al. 2011).

5.7 CONCLUSION

The goal of this chapter was to explore the potential of identifying and managing leadership competencies that contribute to safer workplaces. The chapter breaks new theoretical ground because research on identifying and managing safety leadership competencies is very scarce. Therefore, an exemplary safety leadership competency model was developed, based on a general framework of performance in a leader's role (Campbell 2012) and empirical evidence stemming from studies investigating leadership models and associated outcomes in hazardous work settings. Then, potential benefits and shortcomings of competency modeling in the field were discussed. It becomes evident that competency modeling is a powerful technique that has the potential to systematically improve safety leadership behavior when embedded in an organization's management system (e.g., in human resource management systems). Although competency modeling is widespread in today's organizations, this should not hide the fact that there are a lot of concerns. These concerns refer to the problem that leadership effectiveness cannot be described by an additive set of competencies independent of the hierarchical position of a leader, situational contingencies and temporal factors, and the problem to reliably assess and monitor gaps between desired and existing competencies in order to improve or build up competencies in organizational leaders. In addition, deriving competency models from organizational strategies and objectives striving for safety is prone to the difficulty in detecting and acknowledging leader performance effects on safety. Therefore, applying task analysis methods in competency modeling as well as framing the implementation of competency models as an organizational development intervention seems to mitigate some of these concerns criticizing the mere application of data-driven competency management systems. However, there are also some additional obstacles with regard to the development and implementation of safety leadership competency models, which are briefly discussed in the following section and perhaps, in the best case, will stimulate future research on safety competency modeling.

First, it was assumed that defining competencies as behavioral components fosters a better understanding of required (or desired) leadership performance, but, on the other hand, it is prone to problems in determining objectively measurable threshold levels of performance. One potential solution when developing competency models might be to focus on more indirect performance determinants. And indeed, most of the applied competency models in the field involve a mixture of loosely coupled competencies attributed to performance determinants such as knowledge, skills or abilities, and behavior itself (Sanchez and Levine 2007). With regard to safety leadership competencies, the literature search conducted reveals only a few articles targeting more indirect performance determinants in leaders. For instance, Carillo (2002) provides a five-step leadership pathway (i.e., insight, direction, capability, focus, and accountability) guiding a leader's behavior to develop trust, credibility, and competence in his/her followers. Although these dimensions were derived from practical experiences in the field (and factual case histories of leaders), the validity and reliability of assessing these five variables needs to be proven in the future. Another line of research, conducted by Fruhen and colleagues (Fruhen et al. 2014; Fruhen and Flin 2016), offers some initial insights about a potential mindset of successful safety leaders. The authors developed the concept of chronic unease, which comprises pessimism, propensity to worry, vigilance, requisite imagination, and flexible thinking. Again, an empirical validation of these five characteristics remains to be done and, in addition, it is questionable whether this conceptualization corresponds with the developmental goal of competency modeling, because it is open if all of them can be learned and developed.

Another obstacle refers to the integration of safety leadership competencies in "broader" competency models, which presumably also include other, not directly safety related, competencies such as functional, managerial, and core competencies. As mentioned above, the effectiveness of safety leadership competencies is hard to grasp, and compared with other competencies (e.g., technical knowledge) their assessment might lag behind. In addition, the question remains open of how consistent a competency model is when it simultaneously promotes competencies related to production goals as well as safety goals.

And finally, this chapter aimed at highlighting the potential of competency models to improve the safety culture in a given organization; however, the successful integration of a safety leadership competency model also depends on the maturity of the safety culture at the point in time when such a model is introduced. For instance, Parker, Lawrie, and Hudson (2006) differentiate between five levels of safety culture advancement. Organizations at the lowest level (i.e., pathological cultures) do not have a formal system for day-to-day safety checks and employees are obliged to take care of themselves, whereas in proactive cultures (i.e., organizations with the second highest safety cultural maturity) supervisors engage employees in dialogue, do walk-rounds, and internal cross-audits take place involving managers and supervisors (Parker et al. 2006, 558; see Waterman, Chapter 8 in this book). One can easily imagine that the introduction of a safety competency model will provoke different reactions depending on the given culture. Applying organizational development principles to the implementation process might help to foster organizational members' acceptance of such a model; however, the chances of implementing successful competency modeling are clearly higher in cultures with a higher maturity. Thus, it is up to future research to focus more strongly on the initial conditions that allow for and facilitate a sound introduction of competency models in hazardous workplaces.

ACKNOWLEDGMENT

The author is grateful to Diego Meyerhans (BSc) for conducting the literature search and for his support during the whole process of developing the safety leadership competency framework.

REFERENCES

Badri, A., B. Boudreau-Trudel, and A. S. Souissi. 2018. Occupational health and safety in the industry 4.0 era: A cause for major concern? *Safety Science* 109, 403–411.
Baker, J. 2007. The report of the BP U.S. refineries independent safety review panel. www.ogel.org/article.asp?key=2481 (accessed December 12, 2019).
Baran, B. E., and C. W. Scott. 2010. Organizing ambiguity: A grounded theory of leadership and sensemaking within dangerous contexts. *Military Psychology* 22(Suppl. I), 42–69.
Bartram, D. 2005. The Great Eight competencies: A criterion-centric approach to validation. *The Journal of Applied Psychology* 90(6), 1185.
Boyatzis, R. E. 1982. *The Competent Manager: A Model for Effective Performance*. New York, NY: John Wiley & Sons.

Campbell, J. P. 2012. Behavior, performance, and effectiveness in the twenty-first century. In *The Oxford Handbook of Organizational Psychology*, ed. S. W. J. Kozlowski, 159–196, Vol. 1. New York, NY: Oxford University Press.

Campbell, J. P., and B. M. Wiernik. 2015. The modeling and assessment of work performance. *Annual Review of Organizational Psychology and Organizational Behavior* 2(1), 47–74.

Campion, M. A., A. A. Fink, B. J. Ruggeberg, L. Carr, G. M. Phillips, and R. B. Odman. 2011. Doing competencies well: Best practices in competency modeling. *Personnel Psychology* 64(1), 225–262.

Carrillo, R. A. 2002. Safety leadership formula: Trust + credibility × competence = results. *Professional Safety* 47(3), 41.

Chang, S. H., D. F. Chen, and T. C. Wu. 2012. Developing a competency model for safety professionals: Correlations between competency and safety functions. *Journal of Safety Research* 43(5–6), 339–350.

Christian, M. S., J. C. Bradley, J. C. Wallace, and M. J. Burke. 2009. Workplace safety: A meta-analysis of the roles of person and situation factors. *The Journal of Applied Psychology* 94(5), 1103.

Clarivate Analytics. 2018. ISI Web of Science. http://apps.webofknowledge.com (accessed August 28, 2018).

Clarke, S. 2013. Safety leadership: A meta-analytic review of transformational and transactional leadership styles as antecedents of safety behaviours. *Journal of Occupational and Organizational Psychology* 86(1), 22–49.

Conger, J. A., and D. A. Ready. 2004. Rethinking leadership competencies. *Leader to Leader* 32(32), 41–47.

Donovan, S. L., P. M. Salmon, and M. G. Lenne. 2016. Leading with style: A literature review of the influence of safety leadership on performance and outcomes. *Theoretical Issues in Ergonomics Science* 17(4), 423–442.

Draganidis, F., and G. Mentzas. 2006. Competency based management: A review of systems and approaches. *Information Management and Computer Security* 14(1), 51–64.

Eid, J., K. Mearns, G. Larsson, J. C. Laberg, and B. H. Johnsen. 2012. Leadership, psychological capital and safety research: Conceptual issues and future research questions. *Safety Science* 50(1), 55–61.

Flin, R., and S. Yule. 2004. Leadership for safety: Industrial experience. *BMJ Quality and Safety* 13(Suppl. 2), ii45–ii51.

Fruhen, L. S., R. H. Flin, and R. McLeod. 2014. Chronic unease for safety in managers: A conceptualisation. *Journal of Risk Research* 17(8), 969–979.

Fruhen, L. S., and R. Flin. 2016. "Chronic unease" for safety in senior managers: An interview study of its components, behaviours and consequences. *Journal of Risk Research* 19(5), 645–663.

Guldenmund, F. W. 2007. The use of questionnaires in safety culture research – An evaluation. *Safety Science* 45(6), 723–743.

Hamel, G., and C. K. Prahalad. 1994. *Competing for the Future*. Harvard Business Review. Boston, MA: Harvard Business School.

Hofmann, D. A., and F. P. Morgeson. 1999. Safety-related behavior as a social exchange: The role of perceived organizational support and leader – Member exchange. *Journal of Applied Psychology* 84(2), 286–296.

Hollenbeck, G. P., M. W. McCall, and R. F. Silzer. 2006. Leadership competency models. *The Leadership Quarterly* 17(4), 398–413.

Kurz, R., and D. Bartram. 2002. Competency and individual performance: Modelling the world of work. In *Organizational Effectiveness: The Role of Psychology*, eds. I.T. Robertson, M. Callinan, and D. Bartram, 227–258. Hoboken, NJ: John Wiley & Sons.

Lievens, F., J. I. Sanchez, and W. De Corte. 2004. Easing the inferential leap in competency modelling: The effects of task-related information and subject matter expertise. *Personnel Psychology* 57(4), 881–904.

Lievens, F., J. I. Sanchez, D. Bartram, and A. Brown. 2010. Lack of consensus among competency ratings of the same occupation: Noise or substance? *The Journal of Applied Psychology* 95(3), 562.

Lucia, A. D., and R. Lepsinger 1999. *Art & Science of Competency Models*. San Francisco, CA: Jossey-Bass.

Martínez-Córcoles, M., M. Schöbel, F. J. Gracia, I. Tomás, and J. M. Peiró. 2012. Linking empowering leadership to safety participation in nuclear power plants: A structural equation model. *Journal of Safety Research* 43(3), 215–221.

Martínez-Córcoles, M., F. J. Gracia, I. Tomás, J. M. Peiró, and M. Schöbel. 2013. Empowering team leadership and safety performance in nuclear power plants: A multilevel approach. *Safety Science* 51(1), 293–301.

Martínez-Córcoles, M. 2018. High reliability leadership: A conceptual framework. *Journal of Contingencies and Crisis Management* 26(2), 237–246.

Ozcelik, G., and M. Ferman. 2006. Competency approach to human resource management outcomes and contributions in a Turkish cultural context. *Human Resource Development Review* 5(1), 72–91.

Petitta, L., T. M. Probst, C. Barbaranelli, and V. Ghezzi. 2017. Disentangling the roles of safety climate and safety culture: Multi-level effects on the relationship between supervisor enforcement and safety compliance. *Accident Analysis and Prevention* 99(A), 77–89.

Prifti, L., M. Knigge, H. Kienegger, and H. Krcmar 2017. *A Competency Model for "Industrie 4.0" Employees*. 13th International Conference on Wirtschaftsinformatik, February 12–15, 2017, St. Gallen, Switzerland.

Reason, J. 1990. *Human Error*. Cambridge [England], New York, NY: Cambridge University Press.

Rochlin, G. I. 1999. Safe operation as a social construct. *Ergonomics* 42(11), 1549–1560.

Rodriguez, D., R. Patel, A. Bright, D. Gregory, and M. K. Gowing. 2002. Developing competency models to promote integrated human resource practices. *Human Resource Management* 41(3), 309–324.

Sanchez, J. I., and E. L. Levine. 2009. What is (or should be) the difference between competency modeling and traditional job analysis? *Human Resource Management Review* 19(2), 53–63.

Sanchez, J. I., and E. L. Levine. 2012. The rise and fall of job analysis and the future of work analysis. *Annual Review of Psychology* 63, 397–425.

Sexton, J. B., P. J. Sharek, E. J. Thomas, J. B. Gould, C. C. Nisbet, A. B. Amspoker, et al. 2014. Exposure to Leadership WalkRounds in neonatal intensive care units is associated with a better patient safety culture and less caregiver burnout. *BMJ Quality and Safety* 23(10), 814–822.

Sherman, R. O., M. Bishop, T. Eggenberger, and R. Karden. 2007. Development of a leadership competency model. *The Journal of Nursing Administration* 37(2), 85–94.

Schippmann, J. S., R. A. Ash, L. Carr, B. Hesketh, K. Pearlman, M. Battista, et al. 2000. The practice of competency modeling. *Personnel Psychology* 53(3), 703–739.

Stevens, G. W. 2013. A critical review of the science and practice of competency modeling. *Human Resource Development Review* 12(1), 86–107.

Ulrich, D., and N. Smallwood. 2004. Capitalizing on capabilities. *Harvard Business Review* June, 119–128.

Weick, K. E. 1987. Organizational culture as a source of high reliability. *California Management Review* 29(2), 112–127.

Weick, K. E., and K. M. Sutcliffe 2007. *Managing the Unexpected: Resilient Performance in an Age of Uncertainity.* San Francisco, CA: Jossey-Bass.

Wu, T. C. 2005. The validity and reliability of safety leadership scale in universities of Taiwan. *International Journal of Technology and Engineering Education* 2(1), 27–42.

Yukl, G. A. 2013. *Leadership in Organizations,* 8th ed. Upper Saddle River, NJ: Prentice Hall.

6 Cultural Intelligence
A Construct to Improve Occupational Safety and Health in the Face of Globalization and Worker Mobility Across National Borders

Patrick L. Yorio, Jason Edwards,
Dick Hoeneveld, and Emily J. Haas

CONTENTS

6.1   INTRODUCTION

Globalization and workforce mobility across national borders are increasing trends in today's world economy. These trends have led to increasing numbers of both expatriate occupational safety and health (OSH) managers, and local workforces being comprised of individuals from a diverse array of national cultures. As such, OSH managers are increasingly called upon to facilitate injury and illness prevention initiatives in foreign countries or among diverse groups of individuals socialized in distinct national cultures. Both contexts present challenges and OSH managers must find ways to lead diverse groups of people towards meaningful, safe, and healthy work. Practitioners and academics both agree and advocate that cross-cultural competence is a necessity for managers functioning in today's global economy (Chao and Moon 2005; Ng et al. 2009; Eisenberg et al. 2013). A recent survey from the Economist Intelligence Unit (2012) found that 90% of business leaders surveyed referenced management and communication as a top challenge in their work across national borders, indicating a gap in our current global work environments.

Within this chapter we posit cultural intelligence as increasingly important for organizational OSH in cross-cultural contexts. Cultural intelligence is a prominent model of cultural competence and refers to an individual's capability to effectively manage themselves and others in cross-cultural situations and environments (Earley and Ang 2003; Thomas and Inkson 2004; Leung et al. 2014). As will be discussed, numerous studies outside of the OSH discipline demonstrate that cultural intelligence can benefit a range of important business outcomes in both expatriate and culturally diverse contexts. Further, cultural intelligence can be learned and measured and, therefore, also successfully applied in organizational leadership development initiatives with equal application in empirical research contexts (Caligiuri and Tarique 2012).

In this chapter, we first present some of the challenges that increasing globalization and worker mobility across national borders present to OSH. We then highlight the importance of cultural intelligence in cross-cultural contexts and elaborate on its core dimensions. We then provide a set of examples of how cultural intelligence may be applied to the context of cross-cultural OSH management. We conclude by offering our thoughts on directions for future research in cross-cultural OSH management.

6.2 INCREASING GLOBALIZATION AND WORKER MOBILITY ACROSS NATIONAL BORDERS

Over the last few decades there has been a global increase in foreign workers (World Bank 2018), leading to workplaces being populated by workers with different cultural backgrounds. Recent estimates from the Organization for Economic Cooperation and Development (OECD) suggest that 13% of the total population of their member countries were foreign-born, a substantially higher estimate than the 2000 estimate of 9.5% (OECD 2018). Of these foreign-born, 48% were living in European Union (EU) or European Free Trade Association (EFTA) countries and 34% in the United States (OECD 2018). Somewhat consistent with these figures, in 2017, the International Labour Organization (ILO) estimated that international migrants comprised 18.5% of workers in high income countries, with the highest rates being in the subregions of the Arab states (40.8%), North America (20.6%), and Northern, Southern, and Western Europe (17.8%) (ILO 2018). Further, in the United States (US) the Pew Research Center estimated that in 2050, immigrants will make up approximately 23% of working-age adults, a substantial increase from the 2005 estimate of 15% (Passel and Cohn 2008). Taken together, these reports suggest that increasing workforce heterogeneity based on national culture is a global phenomenon, the trend of which is expected to increase into the foreseeable future.

While some migration reflected within these figures may be seasonal, or short term, many migrate to countries and take on permanent positions. In these situations, training, human capital investment, and/or long-term social and professional relationships may be required and, in turn, the process of integration and labor market assimilation requires investment of significant resources on behalf of an organization (Sparrow et al. 2017).

The unique challenges and opportunities globalization presents to the workforce are gaining increased attention from an OSH perspective as well. For example, although industry-level employment estimates are sparse, some evidence suggests that migrant workers may be more prevalent in high-risk occupations. One report from the UK revealed that the highest proportions of non-UK nationals were working in transport and communications, the wholesale and retail trade, hotels and restaurants, manufacturing, and the agriculture, forestry, and fishing sectors (Office for National Statistics 2017). Statistics such as this suggest that migrant workers appear more prevalent in high-risk industries, a situation which presents unique challenges to OSH. From a worker perspective, the process of integrating into a new culture while, at the same time, working in a high-risk occupation may conceivably be a daunting one. Adapting to a new work environment, learning and adjusting to new safety rules and regulations, creating a new social and cultural life, and overcoming linguistic barriers, all take time, effort, and personal financial resources.

The change in heterogeneity of an employee workforce may add to the resilience and thinking power of an organization (Kochan et al. 2003); however, they can present initial challenges to OSH due to communication and cohesion barriers among team members, possible morale challenges, and different interpretations of the role of leadership and incoming OSH information (Benton 2005; Neuliep 2017; Guldenmund et al. 2013). In addition, immigrant workers may have different perceptions of their physical environment and how to safely and healthfully interact with it (Berry et al. 2002). This could be why a recent Census of Fatal Occupational Injuries within the US workforce found differences in occupational fatality trajectories overtime between groups of workers from different national backgrounds (2013). Within this analysis, a subset of foreign-born workers was the only group to show an increase in workplace fatalities.

Given that workers within an organization are/were also embedded within the broader national culture, globalization and migration present organizational challenges due to culture diversity. As worker values, norms, beliefs, and behavioral tendencies are likely to reflect those of their national culture(s), national culture diversity may significantly influence the barriers raised above, particularly in the case of different perceptions and interpretations of the physical environment, incoming information, the role of leadership, and what is considered the most appropriate way to interact in the workplace. Consistent with these assertions, the Economist Intelligence Unit (2012) survey found that misunderstandings attributed to "differences in cultural traditions" and "differences in workplace norms" to be the greatest obstacle to cross-border collaboration. Thus, culture plays an important role in organizational outcomes in the context of migration and international management and can have a strong influence on the behavioral tendencies of workers. In order to maximize the effectiveness of OSH initiatives, this influence needs to be both acknowledged and taken into account. It is with this recognition that organizational and managerial cross-cultural competence finds its importance.

In lieu of this recognition, and the corresponding substantial implications of globalization and migration for organizations worldwide, within the last decade workplace *cross-cultural competences* have been increasingly identified as a necessity (Ng et al. 2009; Eisenberg et al. 2013; Leung et al. 2014). *Cultural competence* is an umbrella term for the many models and concepts related to intercultural effectiveness independent of the particular intersection of cultures (Ang et al. 2015). Recent reviews of the cultural competence literature have identified more than 300 cultural competence concepts including personality traits (i.e., what a person typically does in intercultural contexts), attitudes and world views (i.e., how a person perceives and evaluates experiences with other cultures), and capabilities (i.e., what a person is able to do to be effective in intercultural contexts) (Holt and Seki 2012; Leung et al. 2014; Ang et al. 2015). *Cultural capabilities* are theorized to be the salient predictor of performance in cross-cultural contexts and a mediator of the effects of personality, attitudes, and world views (Ang et al. 2015). Within this overarching framework, cultural intelligence is posited as a type of human intelligence (Thomas et al. 2015) and "as a theoretically coherent and parsimonious framework of intercultural capabilities" (Ang et al. 2015, 434), thereby emphasizing its potential importance in the process of cross-cultural OSH.

6.3 THEORETICAL REFLECTIONS ON CULTURAL INTELLIGENCE

Human intelligence has been defined as abilities necessary for adaptation to, as well as the selection and shaping of an environmental context (Sternberg 1997). Cultural intelligence, a form of human intelligence, has been described as an individual's knowledge, skills, abilities, and capability to effectively manage themselves and others in cross-cultural situations and environments (Earley and Ang 2003; Thomas and Inkson 2004). Cultural intelligence is focused on the abilities to grasp, reason, and behave effectively in culturally diverse situations (Ang et al. 2007).

Numerous studies outside of the OSH discipline have shown that cultural intelligence is associated with positive outcomes in cross-cultural occupational contexts (Rockstuhl and Van Dyne 2018), including higher quality and more accurate judgment and decision-making (Ang et al. 2007; Ang et al. 2015); enhanced psychological wellbeing, greater satisfaction and commitment while working in foreign national cultures (Berry et al. 2006; Chen et al. 2011; Wu and Ang 2011; Huff 2013); task and group performance (Kraimer and Wayne 2004; Groves and Feyerherm 2011; Crotty and Brett 2012); and effective leadership (Groves and Feyerherm 2011; Rockstuhl et al. 2011).

Research has demonstrated that cultural intelligence is learnable, which lends itself to professional development initiatives in an organizational setting (Earley and Peterson 2004; Ng et al. 2009; Eisenberg et al. 2013; Mor et al. 2013). For example, research in the healthcare industry has shown that cultural intelligence among healthcare professionals is partly acquired through leadership, indicating that leadership-based initiatives should be considered when trying to improve culturally competent strategies (Dauvrin and Lorant 2015).

Further, psychometric measurement devices have been developed thereby increasing its application in a research setting (Ang et al. 2007; Thomas et al. 2012; Ang and Van Dyne 2015; Thomas et al. 2015).

Drawing on Sternberg and Detterman's (1986) view of intelligence, cultural intelligence is theorized to be a holistic capability comprised of four unique dimensions: metacognitive, cognitive (knowledge), motivational, and behavioral (Earley and Ang 2003; Ang and Van Dyne 2015; Rockstuhl and Van Dyne 2018). Although competing theories of cultural intelligence limit its dimensions to cognitive/knowledge, skills, and metacognition (e.g., Thomas et al. 2008), for the purpose of an initial application of the concept to OSH, a broad discussion of the four-factor model is presented.

The *motivational component* has been defined as the capability to direct and sustain energy and focus applied towards learning about different cultures and functioning in cross-cultural situations (Ang et al. 2007; Leung et al. 2014). Given the theorized link between cognition and motivation, Leung et al. (2014) argue that motivation affects whether, and to what extent, an individual directs energy to learn about and understand culturally different others accurately. A lack of individual motivation to intelligently and positively interact with people from different cultures is, therefore, a barrier in the mastery of the remaining forms of the concept. Thus, individuals that are not interested in, or unwilling to learn and sustain culturally intelligent social interactions, will likely not be good candidates to lead cross-cultural organizational initiatives.

The knowledge and behavioral dimensions of cultural intelligence deal with what an individual *knows* about another culture and what they *can do* with that knowledge in order to produce a desired result in cross-cultural contexts (Thomas et al. 2008). *Cultural knowledge* consists of both declarative, content specific knowledge (i.e., recognizing the existence of other cultures and defining the nature of those differences) as well as tacit or process specific knowledge (i.e., intelligence originating from social experience with culturally different others) (Thomas et al. 2015). It consists of an individual's actual knowledge of the specific norms, practices, values, and customs in different cultures and how it likely influences behavior and interactions (Ang et al. 2007; Thomas et al. 2008). This type of knowledge, which includes an understanding of the economic, legal, and social systems as well as the values and norms emphasized within different cultures (Ang et al. 2007), is critical as it provides the basis for intelligent decision-making in cross-cultural situations (Van Dyne et al. 2009). Thomas et al. (2008) described cultural knowledge as the foundation of cultural intelligence because it is the source for comprehending and decoding the behavior of ourselves as well as others. Knowledge of cultural values, attitudes, beliefs, and norms allows for a more accurate prediction and attribution of others' behavior, both of which lead to more effective cross-cultural interactions. In terms of cultural knowledge, Thomas et al. (2015) elaborated that "recognizing the existence of other cultures and defining the nature of difference between them are indicative of the mental processes that are at the core of a systems definition of intelligence" (ibid., 1101).

The *behavioral form* of cultural intelligence refers to the outward manifestations and overt actions of individuals – it refers to what people do and say rather than what they think (Ang et al. 2007) and results in effective cross-cultural interactions when supported by adequate knowledge and metacognition (Thomas et al. 2008). It encompasses an individual's ability to be behaviorally flexible in cross-cultural situations (Leung et al. 2014; Ang and Van Dyne 2015; Rockstuhl and Van Dyne 2018).

Examples of specific types of behavior that can be adapted include how communication is carried out, the tone used during the course of verbal and written interactions, and the proper display of emotions (Emmerling and Boyatzis 2012; Eisenberg et al. 2013). Cross-cultural behavior also accounts for non-verbal gestures such as recognition of physical space, dress codes, and the recognition of norms related to time management (Hall 1959; Eisenberg et al. 2013). As implied, this aspect of cultural intelligence requires that an individual feels comfortable exhibiting flexibility and adaptability in order to properly execute a wide portfolio of behaviors (Ang et al. 2007).

The ability to turn knowledge into effective cross-cultural interactions is strongly dependent upon the *metacognitive component* of cultural intelligence. Metacognition has been defined as knowledge and control over one's thinking and learning activities (Thomas et al. 2008) and it "involves the ability to consciously and deliberately monitor one's knowledge processes and cognitive and affective states, and also to regulate these states in relation to some goal or objective" (Thomas et al. 2015, 1102). Van Dyne and colleagues (2009) argue that these higher order cognitive processes allow individuals to evaluate and revise their thinking, indicating that an increased sense of awareness, accuracy, and response to such situations can improve over time. Thomas et al. (2015) argue that key features of cultural metacognition include awareness of the cultural context, conscious analysis of the influence of the cultural context, and planning courses of action in different cultural contexts.

6.4 CULTURAL INTELLIGENCE APPLIED TO OSH

Cultural intelligence finds its theoretical uniqueness by systematically defining the interconnected process by which motivation, cultural knowledge, cross-cultural behavior, and cultural metacognitive ability converge to result in effective cross-cultural interactions. A valid source of declarative, content-specific knowledge is included within the long history of value-based, cross-cultural research (e.g., Hofstede 1980; House et al. 2004; Kirkman 2006; Andreassi et al. 2014). These large-scale empirical studies were designed to produce knowledge regarding how societies are similar and different in relation to value-based cultural dimensions (Thomas et al. 2015). This type of content specific knowledge, combined with motivation, metacognitive processes (the awareness of the cultural context, conscious analysis of the influence of the cultural context, and planning courses of action in different cultural contexts), and behavior (perceptive acuity, relational behaviors, adaptability, empathy, and tolerance of uncertainty) may be used to illustrate how the most effective organizational safety and health practices in cross-cultural contexts can be developed and implemented. In what follows, we demonstrate how cultural motivation, knowledge, metacognitive processes, and behavior can interact to foster effective OSH management in cross-cultural contexts.

Grounded in the foundational work of Hofstede (1980), numerous cultural dimensions can be used to operationalize differences in the values, norms, customs, beliefs, and practices of national cultures. Individual and group-level cultural orientations on cultural dimensions such as uncertainty avoidance, power distance, and future orientation have been empirically demonstrated to have important implications for a wide variety of organizational outcomes (House et al. 2004; Kirkman et al. 2006). These dimensions can strongly influence how people think and behave both within society in general and in the workplace. In addition, leadership and managerial strategies that are misaligned and/or inconsistent with these values can be ineffective and ultimately undermine the mission, vision, and purpose of an organization (House et al. 2004; Kirkman et al. 2006; Kirkman et al. 2009; Noort et al. 2016). This understanding has been used outside of the OSH discipline to demonstrate that the effectiveness of a set of consistent management practices or leadership characteristics can vary across national cultures (Schuler et al. 2004; Kirkman et al. 2009; Andreassi 2014).

Taken from what has been learned outside of the OSH domain, these same cultural dimensions can also theoretically influence the process of OSH in cross-cultural contexts. Distinct values and personal beliefs may influence worker perceptions of risk and their interpretation of behavioral expectations stemming from OSH leadership and managerial strategies. As a result, not all OSH strategies may be equally effective across organizations with different cultural compositions. Thus, the application of cultural intelligence to OSH should include an understanding of how individuals socialized in different national cultures may react to distinct policies, procedures, programs, and leadership characteristics included within the injury and illness prevention programs.

The application of cultural intelligence to OSH suggests, therefore, that the strategic content and/or implementation (through leadership and vertical social and task-based interactions) of OSH practices, policies, and procedures should be flexible enough in cross-cultural contexts in order to account for cultural differences, which may otherwise limit their effectiveness. Importantly, this line of reasoning should not be taken to imply that fundamentally important practices (e.g., continuous improvement) should be omitted because its national culture may not readily support such a program. Rather, it should serve as a type of tacit knowledge regarding the types of challenges and possible roadblocks OSH managers may encounter when operating in a cross-cultural context, or that may be faced by immigrants working under their leadership. When OSH managers develop this capability, take the time to understand the unique needs of groups within their cultural context, behaving consistently with those needs, and the group is more likely to trust the person and process, thereby facilitating implementation of important injury and illness prevention initiatives (Rockstuhl and Ng 2008).

Table 6.1 provides a few examples of how culturally based values may impact the effectiveness of possible organizational OSH management strategies. Only three cultural dimensions (i.e., uncertainty avoidance, power distance, and future orientation) and five possible OSH management strategies are considered to illustrate the theoretical process postulated. Uncertainty avoidance related to the extent uncertainty is avoided by relying on established social norms and practices; power distance is related to the extent that members of a collective culture expect that power should be stratified and concentrated at higher levels (House et al. 2004). Future orientation is the extent to which members of a collective engage in future oriented behavior such as planning, investing in the future, and delaying gratification (House et al. 2004).

National cultures that are low in uncertainty avoidance (i.e., the extent to which groups rely on norms, rules, and procedures), may be more likely to rely on workers

TABLE 6.1

The Theoretical Effect of Example Cultural Dimensions on Example OSH Normative Strategies

Example OSH normative strategies	Uncertainty avoidance		Power distance		Future orientation	
	High	Low	High	Low	High	Low
Safe work procedures	+	−	+	−	+	−
Decentralized decision-making, autonomy, worker involvement	−	+	−	+		
Accident investigation/post-task safety reviews/ continuous improvement	−	+			+	−
Transformational leadership	−	+	−	+		
Transactional leadership	+	−	+	−		

to handle challenges rather than impose a set of specific safe rules and procedures. For workgroups comprised of workers socialized in such contexts, managers might successfully rely less on the formalization of strict work rules and more on decentralized decision-making, worker autonomy, and work involvement to resolve safety and health problems that arise. Conversely, for workers socialized in high uncertainty avoidance contexts, work process formalization and strict reliance on safe work procedures may be more appropriate (Hofstede 1980; Andreassi et al. 2014). Given that high uncertainty avoidance cultures tend to rely on established norms, continuous improvement initiatives may also be viewed as too risky. Further, given that an increase in work formalization due to high uncertainty avoidance contexts can lead to reduced communication and interdependency between management and workers (Andreassi et al. 2014), the implications of specific forms of OSH leadership may be important. For example, OSH specific transformational leadership may be a less effective strategy in high uncertainty avoidance contexts given the relational and less formalized relationships advocated between leaders and workers.

Similarly, workers socialized in national cultures high in power distance (the boundaries workers expect between leaders/supervisors and subordinates) may not understand or respond to organizational structures characterized by decentralized decision-making, worker autonomy, and worker involvement approaches (Robert et al. 2000; Brockner et al. 2001). Such workers may more readily defer to management to define procedures and safe work rules and be more amenable to transactional types of leadership strategies characterized by extrinsic forms of motivation such as sanctions and rewards for performance and workplace rules, as opposed to safety-specific transformational leadership characterized by open communication, trust, and worker engagement (Kirkman et al. 2009).

As a final example, workers socialized in national cultures low in future orientation (the degree to which members engage in future oriented thinking and acting such as planning and delaying gratification) may struggle with continuous improvement practices such as accident investigation and post-task risk assessments (Goncalves

Filho et al. 2010; Fang and Wu 2013). In cultures high in future orientation, workers may be more risk averse and possibly more likely to follow safe work procedures. Conversely, those socialized in cultures characterized as low in future orientation may be more likely to engage in instantly gratifying behavior such as forgoing the use of personal protective equipment, skipping safety procedures, and/or moving quickly through a task.

This restricted set of examples provides examples of how cultural motivation and knowledge, cross-cultural behavior, and cultural metacognitive dimensions of cultural intelligence may be applied to the context of OSH in cross-cultural situations to enhance effective managerial choices. They also demonstrate the importance of this application: managers must understand varying approaches to direct and indirect communication to effectively manage OSH in cross-cultural contexts (Hall 1969). They also show that an OSH specific form of cultural intelligence may be developed to benefit OSH worldwide.

6.5   CONCLUSIONS AND DIRECTIONS FOR FUTURE RESEARCH

Amid ever increasing globalization of worker migration, cultural differences pose a genuine challenge to successful OSH management. Worker groups are increasingly comprised of individuals from a range of different cultural backgrounds and strategies which have consistently worked for OSH managers in the past may become less suitable over time. Individuals equipped with cultural intelligence tend to have success across a range of cultural settings and will aid OSH managers to overcome these challenges. The dimensions associated with cultural intelligence may also be measured through developed assessment tools (Ang et al. 2007; Thomas et al. 2012; Ang and Van Dyne 2015; Thomas et al. 2015) and are, therefore, potentially amenable to training and development. OSH managers should view mastery of cultural intelligence as a required ability for future success and incorporate cultural intelligence into their personal and professional development.

In addition, while little work has been done to date regarding cultural intelligence in OSH settings, there are clear indications that its benefits may be applicable to all levels of the workforce. As matrix organizational forms, decentralized decision-making, and self-management are growing trends in OSH (Koukoulaki 2010; see Brendebach, Chapter 4 in this book), all members within an organization must be competent in culturally diverse settings. According to Koukoulaki (2010), one of the most important aspects of group self-management of OSH issues is whether they establish group-oriented working rules such as mutual support, consensual decision-making, and self-organized learning. Normalizing these critical aspects in intercultural groups present novel challenges given heterogeneous expectations around social and task-based interactions and norms (Hofstede 1980). Hence, cultural intelligence may be increasingly important to all workers considering emerging forms of occupational OSH management such as self-management and workers participation. Thus, recent group and organization-level theories of cultural intelligence (Ang and Inkpen 2008; Moon 2010) may be particularly applicable to cross-cultural OSH.

Future research into cross-cultural OSH management theory building and hypothesis development is encouraged. While cultural intelligence in its current form will greatly aid the field of OSH management, there may be a need for further theoretical development of OSH dedicated forms of cultural intelligence in the future. Empirical research should seek to study its predictive validity for successful OSH management across cross-cultural settings. Further, we encourage validation of the suitability of various OSH management strategies across cultures. In order to accomplish this, theoretical elaboration of the process of analyzing potential strengths and weaknesses of various OSH strategies depending on important cultural attributes is a critical initial step.

REFERENCES

Andreassi, J. K., L. Lawter, M. Brockerhoff, and P. J. Rutigliano. 2014. Cultural impact of human resource practices on job satisfaction: A global study across 48 countries. *Cross Cultural Management* 21(1), 55–77.

Ang, S., and A. C. Inkpen. 2008. Cultural intelligence and offshore outsourcing success: A framework of firm-level intercultural capability. *Decision Sciences* 39(3), 337–358.

Ang, S., T. Rockstuhl, and M. Ling Tan. 2015. Cultural intelligence and competencies. *International Encyclopedia of Social and Behavioral Sciences* 2, 433–439.

Ang, S., and L. Van Dyne. 2015. Conceptualization of cultural intelligence: Definition, distinctiveness, and nomological network. In *Handbook of Cultural Intelligence*, eds. S. Ang, and L. Van Dyne, 2nd ed., 21–33. Abingdon, UK: Routledge.

Ang, S., L. Van Dyne, C. Koh, K. Y. Ng, K. Templer, C. Tay, and N. A. Chandrasekar. 2007. Cultural intelligence: Its measurement and effects on cultural judgment and decision making, cultural adaptation and task performance. *Management and Organization Review* 3(3), 335–371.

Benton, G. 2005. Multicultural crews and the culture of globalization. *Proceedings of the International Association of Maritime Universities (IAMU), 6th Annual General Assembly and Conference*, 349–356.

Berry, J. W., J. S. Phinney, D. L. Sam, and P. Vedder. 2006. Immigrant youth: Acculturation, identity, and adaptation. *Applied Psychology* 55(3), 303–332.

Berry, J. W., Y. H. Poortinga, M. H. Segall, and P. R. Dasen. 2002. *Cross-Cultural Psychology: Research and Applications*. Cambridge, UK: Cambridge University Press.

Brockner, J., G. Ackerman, J. Greenberg, M. J. Gelfand, A. Francesco, Z. X. Chen, K. Leung, et al. 2001. Culture and procedural justice: The influence of power distance on reactions to voice. *Journal of Experimental Social Psychology* 37(4), 300–315.

Caligiuri, P., and I. Tarique. 2012. Dynamic cross-cultural competencies and global leadership effectiveness. *Journal of World Business* 47(4), 612–622.

Census of fatal occupational injuries (CFOI). 2013. *Census of Fatal Occupational Injuries Charts, 1992–2017*. www.bls.gov/iif/oshcfoi1.htm (accessed January 29, 2019).

Chao, G. T., and H. A. Moon. 2005. A cultural mosaic: Defining the complexity of culture. *The Journal of Applied Psychology* 90(6), 1128–1140.

Chen, A. S. Y., Y. C. Lin, and A. Sawangpattanakul. 2011. The relationship between cultural intelligence and performance with the mediating effect of culture shock: A case from Philippine laborers in Taiwan. *International Journal of Intercultural Relations* 35(2), 246–258.

Crotty, S. K., and J. M. Brett. 2012. Fusing creativity: Cultural metacognition and teamwork in multicultural teams. *Negotiation and Conflict Management Research* 5(2), 210–234.

Dauvrin, M., and V. Lorant. 2015. Leadership and cultural competence of healthcare professionals: A social network analysis. *Nursing Research* 64(3), 200.

Earley, P. C., and S. Ang. 2003. *Cultural Intelligence: Individual Interactions across Cultures*. Stanford, CA: Stanford University Press.

Earley, P. C., and R. S. Peterson. 2004. The elusive cultural chameleon: Cultural intelligence as a new approach to intercultural training for the global manager. *Academy of Management Learning and Education* 3(1), 100–115.

Economist Intelligence Unit. 2012. *Competing Across Borders: How Cultural and Communication Barriers Affect Business.* New York, NY: The Economist, Economist Intelligence Unit.

Eisenberg, J., H. J. Lee, F. Brück, B. Brenner, M. T. Claes, J. Mironski, and R. Bell. 2013. Can business schools make students culturally competent? Effects of cross-cultural management courses on cultural intelligence. *Academy of Management Learning and Education* 12(4), 603–621.

Emmerling, R. J., and R. E. Boyatzis. 2012. Emotional and social intelligence competencies: Cross cultural implications. *Cross Cultural Management: An International Journal* 19(1), 4–18.

Fang, D., and H. Wu. 2013. Development of a Safety Culture Interaction (SCI) model for construction projects. *Safety Science* 57, 138–149.

Goncalves Filho, A. P., J. C. Silveira Andrade, and M. M. de Oliveira Marinho. 2010. A safety culture maturity model for petrochemical companies in Brazil. *Safety Science* 48(5), 615–624.

Groves, K. S., and A. E. Feyerherm. 2011. Leader cultural intelligence in context: Testing the moderating effects of team cultural diversity on leader and team performance. *Group and Organization Management* 36(5), 535–566.

Guldenmund, F., B. Cleal, and K. Mearns. 2013. An exploratory study of migrant workers and safety in three European countries. *Safety Science* 52, 92–99.

Hall, E. T. 1969. *The hidden dimension. Man's use of space in public and private.* London, UK: Bodley Head; Halls Cultural Factors.

Hall, E. T. 1959. *The Silent Language.* Vol. 3. New York, NY: Doubleday.

Hofstede, G. 1980. *Culture's Consequences: International Differences in Work-Related Values.* Beverly Hills / London: Sage.

Holt, K., and K. Seki. 2012. Global leadership: A developmental shift for everyone. *Industrial and Organizational Psychology* 5(2), 196–215.

House, R. J., P. J. Hanges, M. Javidan, P. W. Dorfman, and V. Gupta. 2004. *Culture, Leadership, and Organizations: The GLOBE Study of 62 Societies.* Thousand Oaks, CA: Sage publications.

Huff, K. C. 2013. Language, cultural intelligence and expatriate success. *Management Research Review* 36(6), 596–612.

ILO – International Labour Organization. 2018. *ILO Global Estimates on International Migrant Workers – Results and Methodology*, 2nd ed. Geneva: International Labour Office.

Kirkman, B. L., G. Chen, J. L. Farh, Z. X. Chen, and K. B. Lowe. 2009. Individual power distance orientation and follower reactions to transformational leaders: A cross-level, cross-cultural examination. *Academy of Management Journal* 52(4), 744–764.

Kirkman, B. L., K. B. Lowe, and C. B. Gibson. 2006. A quarter century of culture's consequences: A review of empirical research incorporating Hofstede's cultural values framework. *Journal of International Business Studies* 37(3), 285–320.

Kochan, T., K. Bezrukova, R. Ely, S. Jackson, A. Joshi, K. Jehn, J. Leonard, et al. 2003. The effects of diversity on business performance: Report of the diversity research network. *Human Resource Management: Published in Cooperation with the School of Business Administration, The University of Michigan and in Alliance with the Society of Human Resources Management* 42(1), 3–21.

Koukoulaki, T. 2010. New trends in work environment–New effects on safety. *Safety Science* 48(8), 936–942.

Kraimer, M. L., and S. J. Wayne. 2004. An examination of perceived organizational support as a multidimensional construct in the context of an expatriate assignment. *Journal of Management* 30(2), 209–237.

Leung, K., S. Ang, and M. L. Tan. 2014. Intercultural competence. *Annual Review of Organizational Psychology and Organizational Behavior* 1(1), 489–519.

Moon, T. 2010. Organizational cultural intelligence: Dynamic capability perspective. *Group and Organization Management* 35(4), 456–493.

Mor, S., M. W. Morris, and J. Joh. 2013. Identifying and training adaptive cross-cultural management skills: The crucial role of cultural metacognition. *Academy of Management Learning and Education* 12(3), 453–475.

Neuliep, J. W. 2017. *Intercultural Communication: A Contextual Approach*, 7th ed. Thousand Oaks, CA: Sage Publications.

Ng, K. Y., L. Van Dyne, and S. Ang. 2009. From experience to experiential learning: Cultural intelligence as a learning capability for global leader development. *Academy of Management Learning and Education* 8(4), 511–526.

Noort, M. C., T. W. Reader, S. Shorrock, and B. Kirwan. 2016. The relationship between national culture and safety culture: Implications for international safety culture assessments. *Journal of Occupational and Organizational Psychology* 89(3), 515–538.

Office for National Statistics. 2017. *International Immigration and the Labour Market, U. K. 2016. The Labour Market Characteristics of UK, EU and non-EU Nationals in the UK Labour.* London: Office for National Statistics.

OECD – Organization for Economic Co-operation and Development. 2018. *International Migration Outlook 2018*. Paris: OECD Publishing.

Passel, J. S., and D'. V. Cohn. 2008. *US Population Projections: 2005-2050.* Washington, DC: Pew Research Center

Robert, C., T. M. Probst, J. J. Martocchio, F. Drasgow, and J. J. Lawler. 2000. Empowerment and continuous improvement in the United States, Mexico, Poland, and India: Predicting fit on the basis of the dimensions of power distance and individualism. *The Journal of Applied Psychology* 85(5), 643.

Rockstuhl, T., and K. Y. Ng. 2008. The effects of cultural intelligence on interpersonal trust in multicultural teams. In *Handbook of Cultural Intelligence*, eds. S. Ang, and L.Van Dyne, 206–220. New York, NY: M.E. Sharpe.

Rockstuhl, T., S. Seiler, S. Ang, L. Van Dyne, and H. Annen. 2011. Beyond general intelligence (IQ) and emotional intelligence (EQ): The role of cultural intelligence (CQ) on cross-border leadership effectiveness in a globalized world. *Journal of Social Issues* 67(4), 825–840.

Rockstuhl, T., and L. Van Dyne. 2018. A bi-factor theory of the four-factor model of cultural intelligence: Meta-analysis and theoretical extensions. *Organizational Behavior and Human Decision Processes* 148, 124–144.

Schuler, R. S., I. Tarique, and S. E. Jackson. 2004. Managing human resources in cross-border alliances. *Advances in Mergers and Acquisitions* 3, 103–129.

Sparrow, P., C. Brewster, and C. Chung. 2017. *Globalizing Human Resource Management.* London: Routledge.

Sternberg, R. J. 1997. The concept of intelligence and its role in lifelong learning and success. *American Psychologist* 52(10), 1030.

Sternberg, R. J., and D. K. Detterman, eds. 1986. *What Is Intelligence? Contemporary Viewpoints on its Nature and Definition.* Norwood, NJ: Ablex Publishing Corporation.

The World Bank. 2018. *Moving for Prosperity: Global Migration and Labor Markets.* Washington, DC: International Bank for Reconstruction and Development/The World Bank. http://www.worldbank.org/en/research/publication/moving-for-prosperity (Acccessed on December 12, 2019).

Thomas, D. C., E. Elron, G. Stahl, B. Z. Ekelund, E. C. Ravlin, J. L. Cerdin, S. Poelmans et al. 2008. Cultural intelligence: Domain and assessment. *International Journal of Cross Cultural Management* 8(2), 123–143.

Thomas, D. C., and K. Inkson. 2004. *Cultural Intelligence: People Skills for Global Business.* San Francisco, CA: Berrett-Koehler.

Thomas, D. C., Y. Liao, Z. Aycan, J.-L. Cerdin, A. A. Pekerti, E. C. Ravlin, G. K. Stahl et al. 2015. Cultural intelligence: A theory-based, short form measure. *Journal of International Business Studies* 46(9), 1099–1118.

Thomas, D. C., G. Stahl, E. C. Ravlin, S. Poelmans, A. Pekerti, M. Maznevski, M. B. Lazarova et al. 2012. Development of the cultural intelligence assessment. In *Advances in Global Leadership*, eds. W. H. Mobley, Y. Wang, and M. Li, 155–178. Bingley, UK: Emerald Group Publishing.

Van Dyne, L., S. Ang, and C. Koh. 2009. Cultural intelligence: Measurement and scale development. In *Contemporary Leadership and Intercultural Competence: Exploring the Cross-Cultural Dynamics within Organizations*, ed. M. A. Moodian, 233–254. Thousand Oaks, CA: Sage Publications, Inc.

Wu, P. C., and S. H. Ang. 2011. The impact of expatriate supporting practices and cultural intelligence on cross-cultural adjustment and performance of expatriates in Singapore. *The International Journal of Human Resource Management* 22(13), 2683–2702.

7 Competencies in Safety and Health That Meet the African Complexity and How to Measure Them

Dingani Moyo

CONTENTS

7.1 INTRODUCTION

At a global level, access to occupational safety and health (OSH) services is very low at below 15% levels according to statistics provided by the World Health Organization (WHO n.d.). Countries like Tanzania have access levels to OSH services below 5% (Mrema et al. 2015). There is an important and urgent need to improve access levels to OSH services. The increasing number of occupational diseases, injuries, and accidents in the workplace present a challenge requiring concerted efforts of improving workplace OSH services. The poor scale, quality, and accessibility of OSH services presents a significant rate-limiting step in raising standards. However, there are other influential factors which together determine the take-up of such services and the implementation of any advice on controlling and minimizing exposures to risks. At least, these include the economic and political context ranging from affordability (e.g., in subsistence agriculture) to the choices of prioritizing growth and/or inward investment over the establishment of enforced minimum regulatory standards.

Global initiatives such as the Vision Zero campaign run by the International Social Security Association (ISSA) are invaluable in the improvement of OSH services (ISSA 2017) because they both raise awareness of risks and precautions, and also encourage the development of higher standards which require such services, among other factors, to be achieved. The African economy is diverse, and, in most countries, it is still semi-mechanized leading to a multiplicity of hazards in the working space. One of the important steps in improving the quality of and access to OSH services is through improving competencies of both OSH practitioners and OSH professionals (ISSA 2017). This chapter will explore the main competencies for both OSH practitioners and OSH professionals that are relevant to the unique African context.

7.2   WHAT DOES COMPETENCE MEAN?

Competence has been defined by the International Network of Safety and Health Practitioner Organizations (INSHPO) as the ability to transfer and apply knowledge and skills to new situations and environments consistently applying knowledge and skills to a standard of performance required in the workplace (INSHPO 2017). *Competency* is also similarly defined by the Health and Safety Executive (HSE), UK, as the combination of training, skills, experience, and knowledge that a person has and their ability to apply them to perform a task safely (HSE n.d.). *Competency standards* are according to the International Labour Organization (ILO) a set of benchmarks that define the skills, knowledge, and attributes that people need to perform a work role (Arbabisarjou et al. 2016; ILO 2016). The common principle across these definitions relates to the application or transfer of knowledge and skills to perform tasks. Hence, competencies in OSH relate to the requisite knowledge and skills that OSH practitioners and professionals should possess in order to perform their roles in protecting the safety and health of workers to the required standards. The following discussion will first address the generic OSH competencies and then further describe how these could be applied in the context of Africa.

7.3   GENERIC COMPETENCIES IN OSH

This section will discuss the generic competencies for both OSH practitioners and professionals that are important in the practice of OSH. The capability framework developed by INSHPO describes the activities, skills, and knowledge that are essential for the practice of OSH.

The INSHPO framework covers seven dimensions of important *activities* that are performed by OSH practitioners and professionals:

1. Systems management approach
2. Organizational culture and its impact on OSH
3. OSH risk management
4. Measurement and evaluation of OSH performance
5. Knowledge management
6. Communication, engagement, and influence
7. Professional and ethical practice

As mentioned earlier, *skill* constitutes an integral component of competence or competency. OSH practitioners and professionals should possess skills to perform their roles effectively. These skills include communication, knowledge management, problem solving and critical thinking, evidence-based practice, teamwork, negotiation and management of conflict, leadership, project management and change management, training, and ethical practice. This skill set will be discussed in relation to the African context in the subsequent sections.

The area of *knowledge* includes the following according to INSHPO (2017):

1. The risk management process that includes hazard identification and risk assessment, risk control, and monitoring and review
2. Occupational safety and health management systems
3. Technical and behavioral disciplines

In the following subchapters, the African background for the competence development of OSH practitioners and professionals will be described (7.4). After that, the focus will be on the necessary competencies to perform the role of an OSH practitioner and professional in Africa and the necessary knowledge to do so (7.5). Finally, the issue of the measurement of the competencies will be raised (7.6).

7.4 WHAT IS UNIQUE ABOUT OSH IN AFRICA?

Africa, the second largest continent in the world, boasts of diverse agricultural activities, which are considered the continents' single most significant economic activity employing over two-thirds of the continents' workforce. Furthermore, Africa is a major producer of minerals and metals and its two most profitable minerals are gold and diamonds (National Geographic Society n.d.). The agricultural and mining sectors are characterized by a multiplicity of hazards. In Africa, most workplaces are mechanized to a limited extent while manual labor constitutes a significant portion of the working environment. This exposes workers to significant and diverse hazards such as exposures to pesticides, ergonomic hazards, and injuries. OSH practitioners and professionals should possess good knowledge and skills in order to perform their duties effectively.

There is great diversity in culture and traditions in Africa that have a bearing on the working environment. The diverse African culture not only varies from country to country but within each country as well (Victoria Falls Guide 2018). However, there is a common Xhosa proverb that is common to all African cultures and languages that says, "Umuntu ngumuntu ngabantu" translated in English as, "A person is a person through other people" (Victoria Falls Guide 2018). Cultures have a direct influence on how people behave and perceive issues at the workplace. For these reasons, it is imperative that OSH practitioners and professionals have a good understanding of the different cultures in the countries they work in.

Most African countries lack organized OSH services to cater for the OSH needs of employees. France Ncube and Artwell Kanda (2018) identified the following inadequacies in the organization and practice of OSH in Southern Africa:

1. Lack of some critical categories of OSH practitioners
2. Lack of emphasis on the surveillance of the work environment
3. Disregard of the workers' right to refuse to work in unsafe work environments
4. Non-coverage of some sectors of the economy

The Southern African Development Community (SADC) currently faces varying challenges in OSH service provision that include: OSH human resource capital deficits, lack of comprehensive national OSH systems, and the emergence of an unregulated informal economy which presents significant challenges for the effective implementation of occupational safety and health management systems (OSHMS) (Moyo et al. 2017). In a study conducted in Ethiopia, 89% of the study participants reported a lack of workplace safety and health knowledge or training, while 94% reported a lack of workplace innovation experience (Jilcha and Kitaw 2017). In Tanzania, it was reported that 57% of pesticide applicators who suffered from neurological symptoms cited lack of knowledge in the handling of pesticides (Mwabulambo et al. 2018; see also Elnagdy, Chapter 10 in this book). In another study conducted in Tanzania, 80% of miners who reported higher core body temperatures above the ISO 7933 threshold did not take precautionary measures prior to the commencement of their work (Meshi et al. 2018). This raises important issues like the need for hazard awareness and training. These findings illustrate how constrained the practice of OSH is in some parts of Africa. Inadequate OSH practitioners and professionals negatively impact the practice of OSH thereby creating gaps that are normally closed by non-qualified personnel.

Africa has poorly developed OSHMS as evidenced by the very low levels of access to OSH services and a lack of understanding of foundational concepts such as hazard identification and risk assessment. The emergence of the poorly regulated informal sector in Africa has been associated with poor safety and health standards and environmental degradation (Noetstaller et al. 2004). The informal sector has serious OSH implications as vulnerable populations are indiscriminately involved in this sector. The entry of children, elderly, women, and pregnant women in the informal sector in Africa, calls for stimulating action to better manage the risks through emphasis on preventive actions at the workplace.

There is a significant deficit of trained occupational medical practitioners and occupational health nurses in Africa (Moyo et al. 2015). There is need to develop the necessary competencies of these cadres in order to effectively carry out risk based medical surveillance. This challenge also affects OSH practitioners and professionals who may not have any insight into occupational health (OH) surveillance protocols. This causes problems related to the commitment of resources for carrying out the required OH surveillance. A lack of ergonomists and hygienists presents challenges with regard to workplace design, anticipation, recognition, and evaluation of workplace hazards. Failure to quantify and monitor the effects of these hazards present a challenge to the proper management of workers' safety and health issues. These challenges often stem from the lack of adequate training in safety and health throughout Africa.

7.5 OSH COMPETENCIES TO PERFORM THE ROLE OF AN OSH PRACTITIONER AND PROFESSIONAL IN AFRICA

In the following section, the focus will be on the core activity domains of an OSH practitioner and professional in Africa. For an overview of the required knowledge to carry out these activities see Table 7.1 at the end of this chapter.

7.5.1 COMMUNICATION

Communication plays a significant part in routine OSH practice and constitutes an important competency domain. This is very relevant in the African context where literacy levels are low in countries like Burkina Faso, Chad, Senegal, Sierra Leone, Guinea, Somalia, Chad, and Niger among many others (Post 2015). Generally, safety instructions, material safety data sheets, policies and procedures, and safety signs are predominantly presented in English, French, and other languages. Effective communication coordinates employees, fulfills employee needs, supports knowledge management, and improves decision-making. Soft skills such as communication skills are perceived as important basic skills in safety and health management. Good communication skills are necessary for effective appraisal of hazards and OSH requirements to workers (Blair 2004). The ability to communicate well and with clarity is a key attribute in creating awareness of workplace hazards as well as the general performance of one's duties. At the workplace, good risk communication creates awareness of workplace risks and ensures that workers are aware of the necessary preventive strategies. Clear communication plays a vital role, especially in situations of low literacy levels and naivety in the field of OSH. Managers, supervisors, employees, and OSH practitioners must be good communicators both for instructional purposes and dissemination of safety and health information. Introduction of changes at the workplace requires good communication skills. Furthermore, it is recommended by ILO (2001) that OSHMS should have arrangements and procedures in place to ensure internal and external communication of OSH related issues.

7.5.2 MANAGEMENT AND LEADERSHIP

Key authorities in the OSH field state that leadership in OSH is one of the key elements in the management of OSH (ILO 2001; INSHPO 2017; ISSA 2017). It has also been noted that safety leadership contributes significantly to enhanced safety performance in organizations (Molnar et al. 2019). Implementation and management of OSH in organizations requires competent managers, supervisors, OSH practitioners and professionals, OSH representatives, and workers unions. Good managerial and leadership skills are pivotal in effectively articulating one's roles in OSH. In Africa, it is very common for OSH practitioners to oversee multiple elements of OSHMS that include safety, health, environment, and quality management. Furthermore, the practice of OSH in Africa is often not very specialized and organizations frequently have OSH generalist practitioners whose responsibilities spread across all these disciplines. Managing workers and managers in OSH related functions across different departments in an organization requires good leadership skills.

OSH practitioners and professionals should have good knowledge and skills in management and leadership. Managers expedite certain organizational activities, figure out ways of solving technical problems, serve as peace makers when tensions rise, and make appropriate trade-offs among time, cost, and scope in the routine operations of the organization (Gray and Larson 2003). Understanding these roles played by managers is important in the development, implementation, and integration and monitoring of OSHMS. Zimbabwean companies such as Schweppes Private Limited, Zimplats Mining Company, Delta Cooperation Limited, and ZIMASCO Private Limited, just to name a few, have comprehensive OSHMS due to their good management systems that ingrain OSH into all their business processes.

The practice of OSH requires that goals and actions are established in advance so that an effective and efficient OSHMS can be realized. Through the close interaction with company managers and workers, OSH practitioners and professionals can use their managerial competencies to facilitate the development of corporate OSH policies and procedures. This requires good skills in advocacy, negotiations, and team dynamics in order to influence management and workers into prioritizing OSH issues and ensuring that they are built into the mainstream organizational processes (McShane and von Glinow 2005). OSH practitioners have to work with senior management, line management, supervisors, and workers to ensure a safe working environment within the organization. This also involves educating management in OSH issues so that they are prioritized in the organization. Management has to be accountable for OSH issues at the workplace. As such, all those charged with OSH responsibilities should possess the skills of working with and through workers and managers for the improvement of safety and health issues at the workplace.

Pivotal in the effective practice and management of OSH is competency in human resource management by those charged with this responsibility. Stone (2002) asserts that human resource management involves the productive use of people in achieving the organization's strategic business objectives and the satisfaction of individual employee needs. The purpose of human resource planning is to deploy the resources as effectively as possible where they are needed in order to ensure the effective accomplishment of organizational goals (Muller et al. 2011). OSH practitioners and professionals should possess good human resource management skills. In developing countries, earlier appointments of responsible OSH practitioners and professionals were and are still, in some workplaces, not based on OSH training but on availability. In such circumstances, it is important that such appointments have to be developed in the field of human resource management. Basic skills in recruitment, employee selection, training, and education, as well as dealing with employee relationships are key attributes for OSH practitioners and professionals. Working with employees and employers requires a good ability in dealing with people over and above the OSH technical knowledge. OSH practice can be efficiently articulated by practitioners with good skills in working with and managing people. A good understanding of team processes and the ability to work within teams of different categories are key factors in OSH. In Africa, team building processes are quite often constrained by a lack of financial resources and coaching experts. As a result of these constraints, most organizations in Africa have gained competitive advantages through multi-skilling and multi-tasking in different job categories.

7.5.3 CHANGE

Change management has been noted as a critical element which requires good skills for the effective introduction of OSH changes at the workplace (ILO 2001). For example, developing OSH culture in Nigeria, introducing new OSH programs, and pursuing a particular OSH accreditation standard for OSH practitioners and professionals at national and pan-African level requires good skills in change management. Management of change is part of safety management and may involve technical and organizational aspects (Marko 2017). It may include the introduction of new systems and technologies at the workplace that may be associated with new hazards. This may entail doing business in a different way altogether with regards to OSH principles. Under such circumstances, change has to be well managed to ensure a safe and healthy working environment. This is more important especially in environments where OSH practice is still at its infancy stage or where there are significant resource constraints as is the case in most African countries. The infusion of quality principles into OSH organizational structures and processes requires astute managers and OSH practitioners to maneuver their influence within organizational structures. The African continent still has poorly developed OSHSMs and poorly resourced organizations with regard to OSH human resources. In order to turn around these circumstances and pursue the Vision Zero philosophy, a radical shift in the culture and business processes needs to take effect. Under such circumstances, sound leadership skills are required to offer a catalytic change in the field of OSH. Leadership is that element of management that squarely addresses coping with change. To introduce a new order of work, culture, and systems in OSH, the leadership principle is a key ingredient in articulating a change management process. Change is often resisted and an unwelcome process in most organizations. However, in order to change incongruent safety and health behaviors in organizations and usher in a new dispensation of good OSH practice, it is necessary to go through the change management process. The framework of OSHMS highlights the need for a hazard identification and risk assessment process to be carried out before any changes are made at the workplace (ILO 2001).

7.5.4 CULTURE

Understanding organizational culture and working within different cultures in various workplaces is a key competence area in OSH. The ILO calls for the promotion of a culture of prevention at all workplaces (ILO 2003). The wave of globalization has brought various cultural mixes at workplaces. In Africa, traditional beliefs play a major role in the behaviors of people. Even at workplaces, such beliefs still command considerable influence, in the spirit of "ubuntu" (Victoria Falls Guide 2018). It is important for OSH practitioners and professionals to understand the influence of various cultural dimensions and be able to deal with them in the routine management of safety and health issues at the workplace (see Yorio et al., Chapter 6 in this book).

Organizational culture or corporate culture relates to the basic patterns of shared assumptions, values, and beliefs governing the way employees within an organization think about and act on problems and opportunities (McShane and von Glinow 2005). It is known that organizations that possess a good safety culture experience a reduction in work related risks (Yangho et al. 2016).

Robbins et al. 2001 note that some of the key roles of culture in organizations include the following:

- It conveys a sense of identity for organization members
- Culture creates a collaborative identity beyond an individual
- It builds social stability
- Culture harnesses and solidifies the attitudes and behavior of employees
- It creates distinction between one organization and the other

Artifacts of a good organizational culture such as good safety signage, good physical infrastructure, comprehensive housekeeping, safety slogans, and safety language as well as safety-oriented stories and legends and rituals should be self-evident. Being skilled in the articulation of safety-oriented stories and rituals is important in organizations with diverse cultures.

Due to the diversity of organizational OSH cultures, it is pertinent that new employees are getting familiar with the common, shared norms, values, and behaviors of good OSH practice. There is a high demand of a good skill set in managing OSH in diverse cultures in Africa with its five regions and diverse languages.

Sometimes, new workers from different organizations experience a culture shock which can affect the performance of the OSH system which could even lead to an accident. A cultural shock or reality shock is a natural psychic disorientation that most people suffer when they move to a culture different from their own (Gray and Larson 2003; McShane and von Glinow 2005). OSH practitioners and professionals require knowledge in this area as well as skills to effectively manage new workers coming from different organizational cultural backgrounds.

7.5.5 AUDITING

Auditing plays a special role in the field of OSH by periodically checking the performance levels of OSHMS (ILO 2001). OSH practitioners and professionals should have both the theoretical and practical expertise in the conduct of safety and health audits in the workplace. This is key in the African context where legislation and its enforcement are usually suboptimal. Considering that auditing has a direct role for improving the safety and health of workers as well as programs design and implementation, it remains critically important and very relevant that practitioners possess good auditing skills. In this case, a good theoretical understanding of the auditing function and the possession of comprehensive skills in practically conducting audits are invaluable. Those who lack knowledge on audits and good auditing skills cannot carry out comprehensive audits in the workplace and this negatively affects closure of audit findings and improvement initiatives.

In Africa, where OSHMS are constrained, it becomes difficult to assist workers and organizations in the improvement of workplace safety. Apart from competency in auditing, OSH practitioners and professionals shall have coaching and mentoring skills.

7.5.6 Ethics

Ethics has been defined as the outcome of reflection on the meaning of the concepts of good and bad and right and wrong, as well as on a range of ideas about what confers value or disvalue on human action (Moodle 2011). Ethical and professional conduct in carrying out one's duties in OSH constitutes a very important competence domain. It is an essential requirement that OSH practitioners and professionals are knowledgeable in this area and also possess clear and well-developed skills in conducting their duties in a professional and ethical manner. The International Commission of Occupational Health (ICOH) Code of Ethics provides invaluable guidance to all those involved in the field of OSH (Kogi et al. 2014). It forms an important resource that guides the ethical practice of OSH for all organizational personnel who are charged with the practice of OSH. The ICOH Code of Ethics describes the duties and obligations of occupational health practitioners and professionals such as advisory roles, knowledge and expertise, development of workplace policies and programs, and emphasizing on prevention and corrective actions, among many other important issues. It further elaborates the foundational ethical behaviors of enabling the effective articulation of these duties and obligations. This includes such issues as competence, integrity and impartiality, professional independence, equity, non-discrimination, and communication, among many others. This forms a very important knowledge area for all those charged with the OSH functions in an organization. In Africa where OSH is often poorly legislated and enforced, corruption could lead to compromises in OSH functions. Having knowledge of the principles and provisions of codes of ethics and professional conduct is essential for all those charged with OSH responsibilities in organizations.

Corruption, abuse of office, and lack of comprehensive OSH legislation are prevalent in most African countries. This is further worsened by low OSH literacy levels of most workers leaving them prone to inhumane occupational exposures at the workplace. Practicing OSH under such circumstances calls for practitioners and professionals who can conduct themselves in a professional and ethical manner.

Integrity is a key competence in the professional conduct of one's duties. Integrity refers to being honest or truthful. The protection of workers' health calls for objective and honest application of basic OSH principles. This ranges from the conduct of risk assessments, workplace inspections, auditing, and giving general and specific advice. Failure to be objective and truthful in the conduct of hazard identification and risk assessments can lead to loss of lives, occupational diseases, or property damage. Non-compliant organizations and managers can exert undue influence on the OSH practitioner who lacks integrity. In such circumstances, the safety and health of workers becomes severely compromised. Taking bribes in order to cover up poor safety and health practices at the workplace could lead to compromising safety standards and systems. It is therefore crucial for workers, management, and OSH practitioners and professionals to ensure that standards of practice are objective and professional.

7.5.7 PROCUREMENT AND CONTRACTING

Procurement and contracting play an important role in the field of OSH. They have been described as being integral elements of OSHMS (ILO 2001). OSH practitioners and professionals should play an important role in ensuring that procurement and contracting procedures incorporate OSH requirements. Integration of OSH requirements in the design and management of contracts plays an important role in OSHMS. It is essential that contractors are subjected to the same standards of OSH with regard to hazard identification and risk assessments, education and training in OSH, and accident and emergency evacuation procedures, among other important OSH elements. It is the role of OSH practitioners and professionals to be knowledgeable and skilled in managing contractor safety and procurement procedures.

7.5.8 TOTAL QUALITY MANAGEMENT

Development of robust OSHMS requires management to embrace and apply the concept of total quality management (TQM) in routine operations. A good understanding and application of total quality management to the practice of OSH is essential. Managers and practitioners charged with OSH responsibilities should have knowledge of drawing parallels between the PDSA cycle and the OSH management principles. Total quality management is focused on the customer and seeks to meet or exceed the customer's expectations through a collaborative approach, scientific approach, and continuous improvement philosophy. It is important that practitioners who are involved in the practice of OSH understand total quality management principles and the Deming cycle and the Plan, Do, Study, Act (PDSA) protocol which is applicable to all the stages of risk management (The W. Edwards Deming Institute n.d.). Quality improvement initiatives and corrective actions are part of quality management and should be implemented in the management of OSH (ILO 2001).

7.5.9 MEASUREMENT

Measuring performance in OSH through the use of positive performance indicators and negative performance indicators is an important area in OSH practice. OSH practitioners should be skilled in the design and implementation and monitoring of proactive performance indicators, i.e., positive performance indicators. For measuring OSH performance, they require a good understanding of the use of reactive and proactive measures of safety performance. Furthermore, they should have a good understanding of the reactive indicators, which are the downstream or trailing indicators.

7.5.10 KNOWLEDGE FOR OSH PRACTITIONERS AND PROFESSIONALS IN AFRICA

The main knowledge areas outlined by INSHPO (2017) and discussed under the framework of OSHMS as described by ILO (2001) and ISSA (2017) are described in Table 7.1.

7.6 MEASURING OSH COMPETENCIES

Effective and efficient OSH practice is dependent on OSH practitioners and professionals with requisite competencies. Success in designing and implementing OSHMS in organizations is dependent on the described competencies. Continual improvement in OSH practice requires baselining of the competencies of OSH practitioners and professionals. It is therefore important to assess the OSH competencies so as to ascertain what skills and knowledge they have and also to identify any gaps. The important thing in conducting this exercise requires identification of the assessment methods and the competence areas. Measuring competencies can be done through a number of ways that include the following methods.

7.6.1 QUESTIONNAIRE BASED SELF ASSESSMENTS

Under this assessment method, questionnaires that cover the major OSH knowledge areas and skill sets can be administered to measure specific competencies. However, it must be noted that questionnaires alone could be biased and not necessarily reflect the actual competencies. It is advisable to use a combination of different assessment methods in order to broadly cover the required competencies.

7.6.2 ADMINISTRATION OF WRITTEN AND PRACTICAL EXAMINATIONS

Examinations that cover theoretical knowledge areas and practical skills can be used to assess OSH competencies. Practical examinations can be administered requiring individuals to conduct such activities like walk-through surveys, accident and incident investigations, or development of OSHMS. It must be borne in mind that good examination performance may not necessarily equate acceptable practical performance

7.6.3 360-DEGREE COMPETENCY ASSESSMENTS

OSH competencies can be measured through 360-degree assessments in organizations. This entails assessments of competencies by peers, supervisors, and subordinates. OSH practitioners or professionals can be rated on specific knowledge and skills areas. This method is useful in identifying competency levels of practitioners and professionals.

TABLE 7.1
Knowledge Areas for OSH Practitioners and Professionals in Africa

Risk management process

OSH practitioners and professionals should have a good understanding of the risk management process. Risk management underpins every OSHMS and is very important in the identification and management of workplace hazards. The core of OSH practice is dependent on the risk management process: the identification of hazards, risk assessment, control of hazards, and monitoring and review (New Zealand Standard & Australia Standard 2001; HSE 2019).

Basic principles of hazards and risks	It is important for OSH practitioners and professionals to be knowledgeable on the basic principles of hazards and risks. They should understand the different hazard identification techniques that range from informal methods to the specialized hazard identification techniques. Informal hazard identification techniques include walk-through assessments, consultation of OSH experts, reviews of OSH literature, and feedback from frontline workers. Knowledge on the principles of risk assessment methods using simple matrix tables is crucial. Possession of knowledge and skills in conducting hazard identification and risk assessment is important. Mining and agriculture are associated with key multiple hazards. In Africa where expertise and knowledge in OSH is limited, practitioners and professionals require basic knowledge and skills and the opportunity to develop their experience to enable them to control these workplace hazards.
	It is very important to understand that hazard identification requires a collaborative approach and as such, possession of good knowledge in team dynamics and good negotiation skills are essential. Hazard identification and risk assessment also requires educating and training of exposed populations. As such, knowledge in the local languages and training skills is crucial for one to effectively carry out risk management processes in diverse workplaces, such as in Africa.
Risk assessment	In developing countries where OSHMS are poorly organized or non-existent, practitioners must be able to use qualitative methods of estimating risk levels. It is important to note that risk assessment must be done in a very simplistic way that should be easily understood by the workers. Risk assessment is the process employed in describing the level of risk of injury, illness, or property damage associated with each identified hazard (HSE n.d.). The risk level is determined by the function of the frequency and duration of exposure to the hazard, the probability that the harm will occur, and the associated consequences. Good knowledge in the workplace or work process is vital in order to ascertain the duration of exposure, frequency of exposure and the severity of the injury or illness that can result. The risk level is established by the use of qualitative and quantitative matrices, which translate the severity, probability, and duration of exposure to a specific rating. Hence, from the qualitative matrix a risk level is assigned to each hazard. Risks are then ranked and attention directed towards the hazards that need to be addressed. OSH practitioners must be knowledgeable and skilled in the use of different risk ranking matrices combining the outcome severity and probability of occurrence.
Control of risks	OSH practice requires a good understanding and ability to control risks at the workplace. Having knowledge on the hierarchy of controls ensures a universal approach to risk management in OSH. Risk control can take the form of risk elimination, substitution, engineering controls, administrative controls, and the use of personal protective equipment. The understanding that the hierarchy of controls

(Continued)

TABLE 7.1 (CONTINUED)
Knowledge Areas for OSH Practitioners and Professionals in Africa

	places elimination as the most effective methodology followed by substitution, engineering controls, administrative controls, and personal protective clothing in that descending order of effectiveness is very important. Knowledge and skills in the correct application of the controls is essential. It is important for practitioners and professionals to understand that regardless of the industry complexity or setting, the hierarchy of controls remains a universal method for controlling hazards. It is important for OSH practitioners to know that the treatment regime of risks depends on the level of risk and the organizational context of risk management.
Monitoring and review of workplace hazards	Having knowledge on the fundamentals of monitoring and review of workplace hazards is important. Experience in the effective application of monitoring and evaluation methods is crucial. Skills to evaluate and monitor the performance of the risk management process at the workplace are essential. The ever-changing nature of workplace hazards calls for on-going evaluation of the OSHMS. The advent of new technology, changes in work processes and design, and the introduction of new staff with different skills and knowledge necessitates monitoring and review of already established controls. In OSH practice, knowledge in monitoring and evaluation of OSHMS through self-inspections, independent inspections, periodical inspections, and testing is vital. Planning is an integral part of monitoring. Planning for the introduction of new machinery, new work procedures, and inclusion of OSH requirements in tendering procedures is an important aspect of OSH practice. Most small scale and medium scale enterprises often lack OSH systems and it is the role of practitioners and professionals to ensure that systems are put in place.

Occupational safety and health management systems

A systematic approach in the practice of OSH requires the development and implementation of OSHMS. This provides a systematic approach in managing workplace risks. OSHMS have been found to improve workplace safety and reduce accidents (Wachter and Yorio 2013). It has been found that accident rates, productivity, and general organizational performance improved following the adoption of an OSHMS (Abad et al. 2013). The practice of OSH in Africa has often been fragmented and non-systematic mainly due to a lack of technical competencies in the field. Possession of knowledge in OSHMS and being skilled in the development of OSHMS is essential for practitioners and professionals to be able to effectively implement systems in their organizations. The integral components of OSHMS include the following: OSH policies, managerial commitment, employee participation, specific program elements such as change management, emergency procedures, and auditing, among many others (ILO 2001; INSHPO 2017; HSE n.d.). Monitoring and review of OSH hazards and continuous improvement are part of OSHMS. OSHMS have also been described as covering the following:

- The health and safety work organization and policy in a company
- The planning process for accident and ill health prevention
- The line management responsibilities and
- The practices, procedures, and resources for developing and implementing, reviewing, and maintaining the occupational safety and health policy (HSA n.d.)

(Continued)

TABLE 7.1 (CONTINUED)
Knowledge Areas for OSH Practitioners and Professionals in Africa

Worker participation	Collaboration and participation by workers at the workplace in OSH programs and activities is a key element of OSHMS (ILO 2003; INSHPO 2017). It has also been reported that worker engagement levels act as mediators between the safety management system and safety performance outcomes (Wachter and Yorio 2013). The advancement and success of the OSH agenda at the workplace requires full commitment and participation by workers. Practitioners need to understand that participative management is the primary link at all levels of quality management as well. The success of any operation or OSH management system calls for the support, involvement, and participation of employees in the field of OSH. It is important for OSH practitioners and professionals to understand the need for employees to buy into the health and safety values of the organization. Consultation and involvement of workers in OSH activities encourages ownership of OSH initiatives together with management. The role of worker participation is an important knowledge domain for practitioners and professionals who should appreciate that workers are an invaluable resource in an organization. They have to be skilled in supporting and leading management and worker engagement efforts across the organization.
Education and training	One of the seven golden rules of Vision Zero is to invest in the training and skills of employees and to ensure that the required knowledge is available at every workplace (ISSA 2017). In developing countries, literacy levels in the workplace may be constrained. Training is an important vehicle of changing behaviors at the workplace as well as improving awareness and control of hazards. Training workers, managers, and union representatives on OSH principles, prevention principles, and legal requirements is crucial in capacity building in an organization (ILO 2014). OSH practitioners and professionals should possess training skills and should have knowledge in OSH training. The basic aims of training include the following:

1. Ensuring that workers have the skills to follow safety procedures and perform their work in a safe manner.
2. Increasing the employees' awareness of hazards in the work environment and ingrain a safety culture in the work environment (Bohle and Quinlan 2000).

There are a number of countries in Africa where the literacy levels are below 50% (Post 2015). Creating awareness of hazards and training of workers in such countries requires one to possess good skills in training. Practicing OSH practitioners and professionals should have knowledge and training skills so as to create awareness on the management of hazards at the workplace. This requires good presentation skills and knowledge of the learning processes. OSH training in organizations must include all workers from the least educated to the highly educated. Workers need to be kept up to date with the current trends in the field of OSH. As espoused by the seven golden rules of vision zero, it is necessary that OSH practitioners and professionals have the requisite knowledge and skills in training workers on the management of workplace hazards (ISSA 2017).

7.6.4 CASE STUDIES

Case studies that test specific competencies can be used to assess practical application of skills and knowledge in specific areas. This assessment methodology can be used in combination with other methods described above.

The skills and knowledge areas that were discussed in this chapter should be evaluated using one or a combination of the different methods described above. For Africa, the use of a modified framework that covers the key basic competency areas should be used. This is due to the fact that OSH is an entirely new field that in most areas is staffed by OSH practitioners who lack the required knowledge and skills. It is crucial to ensure that those involved in the practice of OSH possess the basic core skills. The INSHPO competency framework and the Institute of Occupational Safety and Health (IOSH n.d.) Blueprint are important tools that can be used to evaluate OSH competencies. However, modification of these frameworks should be considered against the background of the infancy stage of OSH in Africa and the specific cultural context. The competencies described in this chapter are a first outline of a pan-African competency matrix for OSH practitioners and professionals.

REFERENCES

Abad, J., E. Lafuente, and J. Vilajosana. 2013. An assessment of the OHSAS 18001 certification process: Objective drivers and consequences on safety performance and labour productivity. *Safety Science* 60, 47–56.

Arbabisarjou, A., S. A. Siadat, R. Hoveida, A. Shahin, and B. E. Zamani. 2016. Managerial competencies for chairpersons: A Delphi study. *International Journal of Humanities and Cultural Studies* 3(1), 1654–1665.

Blair, E. H. 2004. Critical competencies for SH&E managers – Implications for educators. American Society of Safety Engineers. *The Journal of SH&E Research* 1(1), 1–16.

Bohle, P., and M. Quinlan. 2000. *Managing Occupational Health and Safety: A Multidisciplinary Approach*. Melbourne: Macmillian Publishers Australia.

Gray, C. F., and E. W. Larson. 2003. *Project Management. The Managerial Process*. New York: McGraw-Hill / Irwin.

HSA – Health and Safety Authority. n.d. *Safety and Health Management Systems*. https://www.hsa.ie/eng/topics/managing_health_and_safety/safety_and_health_management _systems/ (accessed June 26, 2019).

HSE – Health and Safety Executive. 2019. *Risk Assessment*. http://www.hse.gov.uk/risk/con trolling-risks.htm (accessed June 25, 2019).

HSE – Health and Safety Executive. n.d. *What is Competence?- Competence in Health and Safety*. http://www.hse.gov.uk/competence/what-is-competence.htm (accessed June 28, 2019).

ILO – International Labour Organization. 2001. *Guidelines on Occupational Health and Safety Management Systems*. Geneva: International Labour Organization.

ILO – International Labour Organization. 2003. *Global Strategy on Occupational Safety and Health: Conclusions Adopted by the International Labour Conference at its 91st Session*. Geneva: International Labour Organization.

ILO – International Labour Organization. 2014. *Creating Safe and Healthy Workplaces for All*. G20 Labour and Employment Ministerial Meeting. Melbourne: International Labour Organization.

ILO – International Labour Organization. 2016. *Regional Model Competency Standards.* Bangkok: International Labour Organization.

INSHPO – International Network of Safety and Health Practitioner Organisations. 2017. *The Occupational Health and Safety Professional Capability Framework: A Global Framework for Practice.* Park Ridge, IL: International Network of Safety and Health Practitioner Organisations (INSHPO). http://www.inshpo.org/docs/INSHPO_2017_C apability_Framework_Final.pdf (accessed January 12, 2019).

IOSH – Institution of Occupational Safety and Health. n.d. Transforming safety and health across the world. http://www.iosh.co.uk/ioshblueprint (accessed August 16, 2019).

ISSA – International Social Security Association. 2017. *Vision Zero Guide.* http://visionze ro.global/sites/default/files/2017-12/2-Vision%20Zero%20Guide-Web.pdf (accessed June 25, 2019).

Jilcha, K., and D. Kitaw. 2017. Industrial occupational safety and health innovation for sustainable development. *Engineering Science and Technology, An International Journal* 20(1), 372–380.

Kogi, K., G. Costa, B. Rogers, S. Iavicoli, N. Kawakami, S. Lehtinen, C. Nogueira, et al. 2014. *International Code of Ethics for Occupational Health Professionals.* 3rd ed. Monteporzio Catone (Rome), Italy: ICOH / INAIL. http://www.icohweb.org/site/mu ltimedia/code_of_ethics/code-of-ethics-en.pdf (accessed June 28, 2019).

Marko, G. 2017. Safety change management – A new method for integrated management of organizational and technical changes. *Safety Science* 100, 225–234.

McShane, S. L., and M. A. Von Glinow. 2005. *Organizational Behaviour. Emerging Realities for the Workplace.* New York: McGraw-Hill / Irwin.

Meshi, E. B., S. S. Kishinhi, S. H. Mamuya, and M. G. Rusibamayila. 2018. Thermal exposure and heat illness symptoms among workers in Mara Gold Mine, Tanzania. *Annals of Global Health* 84(3), 360–368.

Molnar, M. M., J. Hellgren, H. Hasson, and S. Tafvelin. 2019. Leading for safety: A question of leadership focus. SH@W. *Safety and Health at Work* 10(2), 180–187.

Moodle, K. 2011. *Medical Ethics, Law and Humans Rights: A South African Perspective.* Pretoria: Van Schaik Publishers.

Moyo, D., M. Zungu, S. Kgalamono, and C. D. Mwila. 2015. Review of occupational health and safety organization in expanding economies: The case of Southern Africa. *Annals of Global Health* 81(4), 495–502.

Moyo, D., M. Zungu, P. Erick, T. Tumoyagae, C. Mwansa, S. Muteti, A. Makhothi, and K. Maribe. 2017. Occupational health and safety in the Southern African development community. *Occupational Medicine* 67(8), 590–592.

Mrema, E. J., A. V. Ngowi, and S. H. D. Mamuya. 2015. Status of occupational health and safety and related challenges in expanding economy of Tanzania. *Annals of Global Health* 81(4), 538–547.

Muller, M., M. Bezuidenhout, and K. Jooste. 2011. *Healthcare Service Management.* 2nd ed. Cape Town: Juta.

Mwabulambo, S. G., E. Mrema, and A. V. Ngowi. 2018. Health symptoms associated with pesticides exposure aming flower and onion pesticide applicators in Arusha region. *Annals of Global Health* 84(3), 369–379.

National Geographic Society. n.d. *Africa: Resources.* http://www.nationalgeographic.org/e ncyclopedia/africa-resources/ (accessed June 25, 2019).

Ncube, F., and A. Kanda. 2018. Commentary on the organisation of occupational health and safety in Southern Africa, the International Labour Organization and policies in general. *Annals of Global Health* 84(3), 500–503.

New Zealand Standard & Australia Standard. 2001. *Occupational Health and Safety Management Systems – General Guidelines on Principles, Systems and Supporting Techniques.* Sydney: Standards Australia International Ltd.

Noetstaller, R., M. Haemskerk, F. Hruschka, and B. Drechsler. 2004. *Program for Improvement to the Profiling of Artisanal and Small-Scale Mining Activities in Africa and the Implementation of Baseline Surveys.* Final Report. Washington, DC: World Bank.

Post, africland. 2015. *Ranking of African Countries by Literacy Rate: Zimbabwe No. 1.* http://www.africlandpost.com/ranking-african-countries-literacy-rate-zimbabwe-1/ (accessed June 26, 2019).

Robbins, S. P., B. Millet, R. Cacioppe, and T. Waters-Marsh. 2001. *Organisational behaviour: Leading and managing in Australia and New Zealand.* French Forest, N.S.W: Pearson Education Australia.

Stone, R. J. 2002. *Human Resource Management.* Australia: John Wiley & Sons Australia.

The W. Edwards Deming Institute. n.d. *PDSA Cycle.* https://deming.org/explore/p-d-s-a (accessed June 30, 2019).

Victoria Falls Guide, African Culture, *Vic Falls Bush Telegraph*, September 9, 2018, https://www.victoriafalls-guide.net/african-culture.html (accessed April 1, 2020).

Wachter, J. K., and P. L. Yorio. 2013. A system of safety management practices and worker engagement for reducing and preventing accidents: An empirical and theoretical investigation. *Accident Analysis and Prevention* 86, 117–130.

WHO – World Health Organization. n.d. *Universal Health Coverage of Workers.* https://www.who.int/occupational_health/activities/universal_health_coverage/en/ (accessed June 28, 2019).

Yangho, K., J. Park, and M. Park. 2016. Creating a culture of prevention in occupational safety and health practice. SH@W. *Safety and Health at Work* 7(2), 89–96.

Section II

Case Studies

8 Management and Leadership at Supervisor Level
The Black Hat Program

Lawrence Waterman OBE

CONTENTS

8.1 INTRODUCTION

Hitherto, there has been an emphasis on the role of senior people in organizations leading and managing workforces to achieve high performance in safety. Leadership is about getting people to understand and believe in your vision and to work with you to achieve your goals, while managing is more about administering and making sure the day-to-day things are happening as they should. However, when work was beginning on the construction program for the Olympic Games held in London in 2012, consideration of the experience of site workers concluded that the leadership directly experienced every day by the workers on site undertaking the work was not from senior staff but from their direct supervisors. This was one of the many discussions that took place in the form of workshops engaging the then quite small teams of construction and project managers with health and safety staff. The site supervisors put their teams to work every shift, gave them instructions, dealt with permits to work and other work control arrangements, and intervened to redirect effort if people were not working according to the agreed safe systems of work. The effective day-to-day leaders of construction work were not found in board rooms but were the on-site first-line supervisors. This chapter explains how a supervisor leadership training program was established and how it linked to reward and recognition and other drivers to motivate and maintain high performance. The Black Hat Program represented the key mechanism for achieving the first Games without a fatality, establishing a new approach to safety across the UK construction sector. The chapter concludes with showing how this has developed in the years since, including a renewed focus on occupational health now including mental wellbeing, as well as accident prevention.

8.2   BACKGROUND

The Olympic Delivery Authority (ODA), formed to construct and maintain the venues, the Olympic Park, and the Athletes' Village for the London 2012 Games, committed itself, once established as a legal entity, to achieving high standards in health and safety. At the time of primarily planning, assembling the land, and designing the works (2005–2007) there was in the safety and health community a lot of discussion about the importance and value of clear leadership. Often this was focused on top management and directors, yet quite simple reviews of the nature of site work quickly revealed that for most workers their experience of leadership came from their supervisors, the first line foremen and women who directed the work on site day-to-day and acted as the link to more senior staff. The ODA, working in partnership with its supply chain, devised and executed a program deliberately to develop those supervisors as leaders. The supervisors were identified as the layer that would be able to achieve the high standards we aspired to and in particular our health and safety targets. Our main aims were being the first Olympic Games to be constructed without a fatality, establishing a new industry norm of very few accidents, and the prevention of work-related ill health (Waterman 2014).

8.2.1   NATURE OF THE LONDON 2012 PROJECT

The London 2012 construction program was initiated when London's bid to host the Olympic Games was declared successful by the International Olympic Committee in Singapore in June 2005. Planning accelerated and practical work on the site began: land assembly, site investigations, and the start of tunneling to underground electricity supply cabling carried across the site on 52 towers. The construction works, funded by the UK Government, commissioned and overseen by the ODA, encompassed the Olympic Park and the venues housed there, which at the time was the largest regeneration project in Europe. Components of this project were the Athletes' Village, Europe's largest new housing project, and other sites remote from the Park, including those that hosted sailing, canoeing, shooting, and a number of training venues – both permanent and temporary. The large site was embedded in an area of east London which by many different criteria was defined as poor, run-down with pockets of significant poverty, and generally low educational attainment, high unemployment, and street crime – a suitable target for regeneration.

The program divided into separate projects, infrastructure including sewerage, roads, and bridges, and venues such as the Velodrome, Aquatics Centre, and the Olympic Stadium, and each project was led by a main construction contractor. Each main, or principal contractor, developed and brought to site its own supply chain of specialists from ground workers for the initial site preparation to electricians and others engaged in the final fit-out, teams arriving and departing as the projects required. As work progressed, the management team of each principal contractor remained in place, but beneath them the site workforce changed as different skills were needed at each stage of construction. Several phases formed the construction schedule: enabling work to prepare the site for development; venue and infrastructure work to build the facilities for the Games; deconstruction of the temporary structures at the end of the Games; and preparation for legacy work to transform the sites and venues for future use. The Olympic Park is now designated the Queen Elizabeth Olympic Park and has received more than 16 million visits since the Games.

Construction was undertaken by principal contractors procured in competition, and the ODA's specifications included a "Health and Safety Standard". This Standard was a set of commitments and requirements that were written into each contractor's formal contract outlining what we wished to achieve and the mechanisms to be employed (UK's National Archive 2012).

Health and safety, linked to security (a strategic decision following the London terrorist attacks which took place the day after London was announced as the venue for the 2012 Games), formed one of a series of six priority themes: equality and diversity; employment and skills; design and accessibility; sustainability and legacy, defining how we work. The ODA employed a delivery partner, CLM (a consortium composed of the companies CH2M Hill, Laing O'Rourke, and Mace) to project manage and oversee the works, including delivery of these priority themes.

8.2.2 CONSTRUCTION WORKFORCE

From a total UK workforce of 29 million, 2.3 million worked in construction during the main build for the 2012 Games (Office for National Statistics 2010). Only 13% of people working in construction were female, biased towards administration, design, engineering, and other "white collar" jobs. People "churn" through such a program – early on there is no requirement for plasterers and electricians as buildings are demolished, ground is cleared, and drains installed. Later, the people engaged in earth movement give way to workers planting for the landscape, and in the buildings, the final fit-out requires a range of skills not used early on. As a result, although the maximum number of workers at peak construction provides one image of the workforce – 12,500 workers in 2010 – overall, 50,000 people were inducted onto site. This level of churn was less than for other major projects, as some workers were able to move from one project to another, and were encouraged to do so as they had become familiar with the journey to work, the working arrangements, and other aspects of the job. This level of churn meant that establishing and maintaining a culture including a health and safety culture, a feeling of "this is the way we work around here", with reliable, repeated behavior patterns was potentially harder to achieve than perhaps applies to more stable working environments with a relatively static workforce, such as is more typical in manufacturing. This is one of the reasons why supervisors were seen as an important link between management and the ever-changing workforce, as this higher skilled layer tended to be "first in, last out" of any team as its totals waxed and waned through its contracted work.

US research sheds light on the construction workforce, data from 2010 indicates that the educational attainment of employees in construction is lower than in most other industries except for agriculture. In 2010, about 40% of construction workers had some post-secondary education, in contrast with 62% of the total workforce (CPS 2010; CPWR 2013). The issues of education, including literacy, had to be taken into account when establishing a communication program for a large workforce. Arguably, much of construction was and is a verbal industry and this influenced the emphasis on "Daily Activity Briefings", which prior to the London 2012 work were not common across the sector, and more generally on the verbal communication skills of supervisors.

The final general comment on the construction workforce is the common lack of connection with health services despite the free-at-the-point-of delivery nature of the UK's National Health Service. Construction has a predominantly male and aging population, according to the 2011 census data the highest proportion of workers in construction were aged 40 to 44 years (13.4%), followed by 45 to 49 years (13.3%), and 35 to 39 years (11.1%). The youngest workers, aged between 16 and 19, accounted for only 2.7% of the total construction workforce in 2011, while 20 to 21 years accounted for 3%. Older males have been found to exhibit a consistent reluctance to present symptoms to doctors in general practice (Courtney 2000), and the UK construction industry had prior to London 2012 an extremely poor record of providing on-site occupational health support. This is compounded by the peripatetic nature of the workforce, with many individuals traveling to and from family homes that could be hundreds of miles away – journeys on Fridays to the family and back to work for Monday morning leave little time to make appointments and see a family doctor, and the on-site health service found that few registered with general practitioners local to their worksite.

There are two further issues relevant to efforts to create full engagement between management and the site workforce which were present during the project and remain applicable. First, in the UK, employers are required to make a financial payment for each employee under the heading National Insurance, and to pay for holidays (entitlement is to a minimum of 5.6 weeks per year), and workers are also entitled to a level of payment if they are away from work on sickness absence (for up to 28 weeks) and for paid maternity or paternity leave. Construction employees also receive free personal protective equipment (PPE) from their employers. However, if a worker is designated as self-employed, none of these entitlements and payments apply. Contractors often encourage their workers to be self-employed with fewer rights, and as the self-employed pay tax later, can limit their national insurance payments and also claim work-related expenses, such as travel, tools, and so on, against their tax liability – many agree to do so. As a result, as many as 300,000 UK workers who are in practice engaged by a single contractor who controls their work as if they were an employee, operate as self-employed, often referred to as "bogus self-employment". The difficulty is that the employers of such workers historically have taken less responsibility for training, management, and control. Many workers have developed in a culture of individualism and partial autonomy that does not align with a disciplined approach to compliance and enacting method statements and other procedures defining safe work (Seely 2018). The second issue is of victimization and blacklisting which several major construction companies engaged in – some considered the workers to be "troublemakers", others that they were activists motivated by a desire to improve workers' conditions–by financing and making use of "The Consulting Association" founded by the major contractor Sir Robert McAlpine plc. Court cases, which are continuing, have so far generated compensation to affected workers of between £50 million and £250 million, and have spurred the UK Government to introduce a legal ban on blacklisting–and the issue has also reached the European Court of Human Rights (Smith and Chamberlain 2015). The importance of the impact immediately on 3,000 workers cannot be overestimated, but it had a much wider effect by encouraging a defensive, sullen silence amongst workers – the opposite, for example, of an engaged workforce reporting near-misses. It is just a short time since this shady practice was aired in the courts and parliaments of Scotland and the UK, and the practice undoubtedly caused a real fear amongst workers that if they spoke up – for example, raising concerns over safety matters – they would end up unemployed and unemployable. Because of the natural lifecycle and churn in construction, people did not have to be sacked, they were simply turned away from opportunities for their next project. In this "verbal" industry, references are often taken up by telephone and a simple word against someone would be enough to discourage another person from employing them. Creating an engaging safety culture amongst the bogus self-employed and those with realistic fears of blacklisting was extremely difficult in the period of the London 2012 construction program.

8.2.3 Accidents and Ill Health History

Construction is a large employment sector and one of the most dangerous. As work got underway for the 2012 London Games, roughly a third of all UK work fatalities happened in construction and the fatal injury rate of 6.0 per 100,000 workers (2000/2001) was over three times the average rate across all industries. The London 2012 program was estimated by ODA project managers to require about 100 million person hours to be worked. A simple analysis indicated that if the works reflected the industry averages, even biasing them towards the lower rate for major projects, it would result in three worksite deaths and about 750 serious injury accidents of which a third would be life-changing (injuries from which there would not be a full physical recovery) (HSE n.d.). The data on health was just as concerning, albeit almost matched by the paucity of reliable information – ill health has historically been poorly identified, recorded, and reported – with estimates of tens of thousands of workers suffering from work-related ill health each year and up to 40,000 of these being new cases. Taking just one aspect, in 2005, just as work was getting underway, about 8,000 occupational cancer deaths were recorded, of which 3,500 were assigned to the construction sector which employed only about 7% of the national workforce. The poor record of construction, addressed by special summits involving business leaders and politicians and regulations targeting poor management in the sector, encouraged many of those involved to seize on the London 2012 program as an opportunity to effect a step change in health and safety management (Prescott 2001).

8.2.4 HEALTH AND SAFETY COMMITMENTS

Working together and in parallel, the *Strategic Forum for Construction* developed and published the "2012 Construction Commitments". Under the headings of procurement and integration, commitment to people, client leadership, sustainability, design quality, and health and safety, it sought to map out the key factors identified to improve the industry that in 2006 represented almost 10% of UK gross domestic product (GDP). The ODA worked up in more detail what it sought from its contractors in its "Health and Safety Standard", which outlined aspects of the management system required but not standardized: from competent workers and training to behavioral safety programs, participation in the national "Considerate Constructors" scheme (CCS n.d.), and incorporation of the "Respect for People" guidelines referenced in the commitments document. Much of this material was based on reports from *Constructing Excellence*, an industry and government supported body which was formed under the impetus of two key reports into construction (Latham 1994; Egan 2014; Constructing Excellence 2015).

There were headline targets, such as zero fatalities, never previously achieved for the base build of an Olympic Games, and requirements to engage with occupational hygiene and health services provided by the ODA through "Park Health" established on the park and within the Athletes' Village sites to address the occupational health needs of day-to-day construction activities and the large workforce. The Standard, for example, in later editions following its launch in 2006, published a red/amber/green list of design considerations largely drawn up for the ODA by a partnership between engineering design companies Arup and Atkins appointed to design the infrastructure of the north and the south of the park respectively. This sought to reduce the use of hazardous substances and dust-producing processes and encourage off-site manufacture and other techniques known to enhance risk control.

8.2.5 HEALTH AND SAFETY PROGRAM

The ODA established a program that stepped through the elements required for a comprehensive approach to health and safety. This is shown in Figure 8.1.

The philosophy was straightforward: establishing the health and safety aims and aspirations of the program and outlining some of the key methods to be employed created a framework documented in the Standard and written into the construction contracts employing the whole supply chain. Each company within that supply chain was then required to employ its management system to embed and work towards those aims. The destination and some of the steps were laid down by the ODA, but the detailed work was undertaken by each company employed on the program and so there was a high degree of variability in style. Some of the construction teams

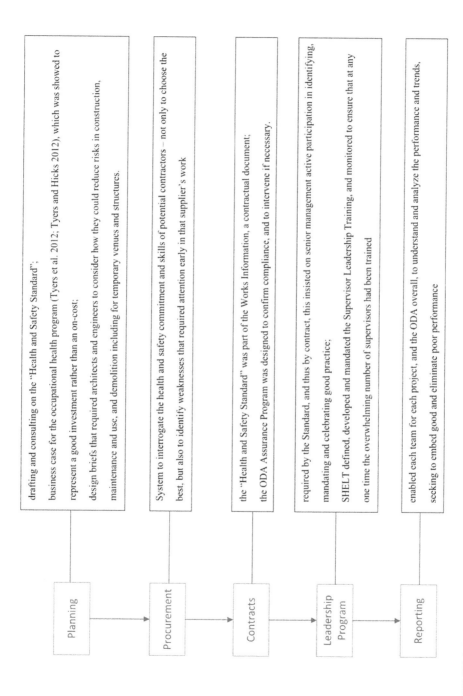

FIGURE 8.1 ODA approach to health and safety.

were led by quite charismatic people, others operated as much more of an integrated team, but all worked coherently within the framework and auditing by the ODA and its delivery partner assured and intervened as necessary. In management, most of the organizations had formal systems based on OHSAS 18001 (now superseded by ISO 45001) or similar to it, but again style varied from the highly consultative to something that looked much more compliance-driven "command and control". This was, perhaps, one area where the ODA approach encouraged more respect for the site workers and considerably more engagement with them in the development rather than just the policing of site work practices (IES 2013).

8.3 COMPETENCE FRAMEWORK

The Olympic Delivery Authority Board and Executive had accepted the "threat" represented by the calculation of a predicted three fatalities during the planned works if the program simply represented "business as usual". From the beginning, the senior people recognized that they needed to act as forceful leaders, to express a collective determination to achieve high performance standards in health and safety, and to mobilize and motivate leadership to operate at the levels below. The first such level was created by bringing together the business/project leaders of the major supply chain companies with representatives of the ODA and its delivery partner.

8.3.1 LEADERSHIP

The leadership program for the London 2012 construction works was populated by the supply chain project directors, one from each of the principal contractors responsible for each project. The directors, with representatives of the ODA and the delivery partner, met monthly as the "Safety, Health and Environment Leadership Team" (SHELT). At each meeting, this powerful grouping, including people responsible for managing projects with a value up to £450 million, discussed recent performance and "news" from anywhere on evidence of good practice. If the leaders agreed that a particular initiative promised improvements, they would adopt that approach. After everyone had agreed and reported that the initiative was up and running on the projects on site, where appropriate, they were then incorporated into a manual of "London 2012 Common Standards". The loop was closed when delivery partners' health and safety staff, during their usual assurance checks, were able to confirm that the standard was common practice across the program of works. On large complex projects there are two forums in which the leaders of individual projects can be brought together productively, one is logistics aimed at organizing the site for efficient operations, managing deliveries and waste, and similar matters of common interest – and the other, if there is client leadership, is on health and safety. In fact, SHELT was the only forum on the Games construction program which brought all the project heads together and created the opportunity to establish a common site culture. It should be noted that it is possible to discuss leadership in terms of style – charismatic, transformational, and so on – but in practice, each of these designators can be found embodied in the people and their actions in a real situation. This chapter focuses on the mechanisms that were effectively employed rather than outlining some of the more superficial aspects of the form.

8.3.2 Focus on Supervisors

Research, as well as anecdotal practical experience, had strongly suggested that the ambitious targets for high health and safety performance in London 2012 would need the engagement and support from site supervisors (Hinzie and Raboud 1988). One of the interesting areas of study has been exploring the potential role of supervisors in developing countries where the overall levels of experience and competence of site workers may be expected to be low: According to my personal experience, while working on three solar power projects in 2016–2017, many workers, for example in Morocco, were seasonal agricultural workers who supplemented their incomes with short periods of work on construction projects. One study looked at the impact of employing a competency framework to supervisor development (Serpell and Ferrada 2007), and the arguments about the particular challenges of an inexperienced construction workforce that could be extended to a competence framework when one considers those issues referred to above concerning education and site culture. However, the London 2012 leadership team did not limit itself to enhancing competence, encouraging the development of instrumental skills, but extended this to explore the qualities of a positive on-site culture that encouraged and expected good communications and respectful relationships between individuals and teams. One perspective was provided in a speech given just after the Games by the former Chief of the General Staff (Head of the British Army) General Lord Dannatt, when he described health and safety as the bedrock on which the program of construction rested. The approach to not just train and develop but also to engage and involve was the major contributor to establishing a site culture of cooperation and protection. SHELT was satisfied that the supervisor was a crucial communication and leadership link in the chain from project office to site and it was SHELT that decided to develop and launch a training program for supervisors as on-site leaders.

8.3.3 Black Hat Training

Stimulated by both evidence and previous experience of the positive role supervisors could play, the leadership team agreed that each project supply chain, from the principal contractor down, was responsible for ensuring that their site supervisors were technically competent. That is, for groundworks, scaffolding, and so on the supervisors of each work team could only be in that control position if they had a good understanding of the risks and appropriate controls for the particular tasks associated with that trade and set of activities. However, the construction industry had not, hitherto, developed supervisors by helping them with clear expectations of what the role entailed and with the skills of leadership and communication. For most workers on most working days, they had little or no contact with managers far less directors, but every day they were set to work and overseen by a supervisor. The supervisors were, clearly, the "leaders" that the workers actually experienced so if the leadership program had simply stuck to the SHELT meetings and issuing standards, it would not have been impactful at site level.

The outcome of these considerations of where and what was the feeling of leadership, or "felt leadership", for site workers was at two levels. First, a program was initiated of raising the supervisors' sense of authority, responsibility, and seniority. They were identified, across all the projects irrespective of employing company, by an obligation to wear black hard hats. Indeed, as a classic example of synecdoche we began to refer to supervisors as "black hats". The leadership team, through each contractors' management, invited the supervisors to program-wide meetings and events we called "black hat conventions", where supervisors from different projects and teams met each other and discussed the work and the health and safety challenges of their roles.

The major development was the establishment of a training program for the supervisors. This was originally drafted by on-site managers working with the Construction Industry Training Board (CITB). The course was very straightforward, seeking to help the attendees:

- understand the significance of their position in helping the program manage risk;
- develop communication skills, so that they could comfortably provide their teams with a short briefing at the beginning of every shift on that day's work, including an outline of any risks and precautions;
- include listening skills, so that, for example, at the "Daily Activity Briefing" they provided a real opportunity for their workers to raise any issues, share any relevant information; and
- recognize and challenge inappropriate behavior and working methods, even when exhibited by experienced workers or anyone with a strong, resistant personality.

The arrangements for supervisors were defined in a common standard, developed and issued as described above.

8.4   COMMON STANDARD FOR THE SUPERVISORS' TRAINING COURSE

The training course requirements for developing supervisors as leaders were later published by the ODA as part of its "Learning Legacy", in the anticipation that it may be of use to future projects seeking a best practice example (IES 2013a; 2016). This was in a series of tools and templates issued as "Champion Products"; tools and formats used by the ODA in executing its program. The mechanism for setting a common approach across the London 2012 construction program was through SHELT. The directors, joined by representatives of the ODA and of its delivery partner, discussed health, safety, and environmental performance and opportunities for making improvements. Suggestions were made that fell into one of three categories:

a. An idea that seemed so positive and beneficial that the contractors agreed to adopt and mandate it following little discussion. An example of this was the arrangement to eliminate hammering into the ground of metal pins for marking out, with the associated risk of hitting subsurface services.

b. An idea that may have been deemed appropriate for one or more of the contractors, but not achieving general support, was that in addition to random drug and alcohol testing across the workforce, some of the contractors implemented such testing on the first day of attendance of all new workers to the site.

c. An idea that would eventually fall into one of the categories above, but as yet there was inadequate information and evidence of the idea's impact. It was agreed to develop the idea and evaluate its effect, but not to set it as an agreed minimum standard that everyone was mandated to follow.

The initiative that was proposed by SHELT, and then developed with the National Construction College, was to train construction supervisors to better engage with the workforce, with an aim of improving behavior and culture for health and safety. It became Common Standard 38, and was used to raise the level of workforce competence and encourage a behavioral shift in supervisors acting as leaders, which provided an effective link between senior site managers and the workforce (UK's National Archive 2018). This was envisioned as a key part of achieving an engaged workforce and was soon adopted by the pan-industry body the UK Construction Group as a new, general standard for the industry (Build UK 2019).

Competent supervision of the workforce is not only a requirement under UK law but was recognized as essential for the safe and timely delivery of construction work to the required quality. The general standard was in effect from November 2009 and required that all supervisors working on the Olympic Park could demonstrate that they had the necessary knowledge and skills to supervise their workers effectively. This was necessary if they were to ensure those workers carried out their tasks safely and with due regard to health. For the purpose of the standard, supervisors were defined as those persons directly supervising work activities. They were the frontline supervisors with direct responsibility for putting people to work and who would typically brief their workers on how to carry out their work and ensure they were carrying out their work safely and healthily. An earlier mandated standard was the provision of the "Daily Activity Briefing" to every worker at the beginning of every shift, and this responsibility largely fell on the supervisors to carry out.

Supervisors were required to have undertaken training that met the technical standards, and the responsibility for this already fell to their employers who had promoted and appointed them into their supervisory positions. Supervisors had to be able to demonstrate that they had sufficient knowledge of health and safety gained at a construction specific course of at least two days duration with the following as a minimum requirement:

- The role of the supervisor
- Health and safety law and enforcement
- Safety systems of work including hazard identification, risk assessment, method statements, and the general principles of prevention including the hierarchy of controls
- Health and safety management including safety policies, measurement performance, and permit systems

- Occupational health including the sources of risk of exposure to noise, vibration, asbestos, dermatitic agents, and asthmagens, the controls and precautions available, and the role of health monitoring and the drugs and alcohol policy
- Fire prevention, control, and other emergencies including types of fire, fire prevention (fire plan), raising the alarm and escape, firefighting, legal requirements
- Hazardous substances including risk assessment, precautions and control, legal requirements
- Duties relating to health and safety, and welfare on construction sites
- Working at height
- PPE including selection and maintenance, information instruction, and training
- Importance of provision of a demonstrable method of recognizing supervisors on site
- Manual handling including definition, manual handling assessment, information, and training
- Accidents and first aid

The Common Standard then homed in on the requirements for supervisors, readily identifiable by their black hard hats (mandated in a general standard on PPE), to be cultural leaders. The program was designed to encourage them to help establish, demonstrate, and maintain the site behaviors which would be able to protect workers from risks of accidents or to their health. In Part 2 of the requirements, the focus was not on technical skills associated with working at height or with hazardous substances, but rather on the supervisory skills to lead a team, issue impactful instructions, challenge inappropriate ways of working, and similar elements of leadership. Supervisors had to be able to demonstrate that they had sufficient skills to be an effective supervisor of their workers. This required them to attend training which included:

- The role of the supervisor
- Understanding behavioral issues
- Leadership and effective intervention skills
- Delivering effective presentations (e.g., "Daily Activity Briefings" and "Tool Box Talks")
- Role-play

While SHELT and the ODA left responsibility for the technical, Part 1 of the training to the employers, with assurance checks to confirm that supervisors were trained and competent in these aspects of their role, the Part 2 training in leadership was a course developed for and run by the ODA on site.

The plan was to raise the competence, and also the motivation and commitment, of the supervisors across the program. It was accompanied by efforts to ensure that there were good ratios of supervisors to workers: exact proportions were not stated. It was understood that during early works on a cleared site, fewer supervisors could keep track of the work of large teams. In contrast, during snagging works on nearly complete buildings, when workers were separated into individual rooms and other workspaces, undertaking multiple tasks in properties that had live services, closer supervision was appropriate.

Both parts of the training required testing to confirm that learning had occurred, and the successful candidates were issued with certificates to demonstrate a suitable standard had been achieved. The first phase of training and certification was already in place, often from training courses recognized by specialist trade associations and bodies, with employers aware that they had a legal duty to ensure the technical competence of their appointed supervisors. The new element was the leadership skills area addressed in Part 2, and here the ODA working with the Construction Training College established a course run on site over 1½ days. Supervisors usually attended a half day in each of three consecutive weeks. Contractors were also advised that their existing in-house courses could be adapted to meet the above required criteria. Generally, holders of a recognized health and safety course certificate which did not match the above were able to "top up" by attending the Part 2 course.

It should be noted that supervisors for whom English was their second language, for example, eastern European workers from EU countries, and therefore eligible to work in the UK at that time, had to be supported by another supervisor able to communicate effectively with them. As each course was run, a representative from SHELT welcomed the delegates and explained why the course was being run. The same person then returned before the end of the third session to thank the attendees, ask them to provide some feedback on the course, and outline anything that senior people such as him/her representing SHELT could do to support their work as supervisors. This powerful endorsement of the course and support for the role of supervisors was cited positively by workers in surveys exploring the safety culture on site. The importance of on site supervisors was celebrated with the help of a poster. (Figure 8.2)

8.5   BLACK HAT CONVENTIONS

Alongside the training and motivation of supervisors individually and in small groups, the ODA encouraged a sense of a leadership layer throughout the program of works across the many projects. One mechanism was holding "Black Hat Conventions", another element of this proleptic approach. We were convinced that if we treated supervisors as leaders, it would encourage them to become leaders. In a local theatre, or later within one of the larger venues, every three months or so, supervisors would come together from across the many projects on site. Initially addressed by specialists and managers on different topics and then encouraged to discuss and debate for up to an hour, soon the conventions were addressed by the supervisors themselves having pre-prepared talks on, for example, the implications of venues going electrically "live". After one hour of business held towards the end of the normal working day, refreshments were served and the supervisors were free to mingle and chat to each other with members of SHELT and other senior managers helping to host and encourage discussion. They were very friendly, positive affairs where the challenges of being supervisors could be discussed "amongst friends". They also created a sense of camaraderie, of a cadre of leaders with a common

FIGURE 8.2 Poster celebrating supervisors "graduating".

interest in achieving the highest health and safety standards. It was discussion at these gatherings that led to the supervisors themselves setting up mentoring for new supervisors by more experienced colleagues. The discussions sometimes identified further training requirements and the training being made available. Generally, the conventions acted as a mechanism for generating pride in and respect for the role. The importance was, above all, in following up on the training with mechanisms for demonstrating respect, providing support, and maintaining motivation.

8.6   DAY-TO-DAY ON SITE

The program to develop supervisors as leaders resulted in many conversations between SHELT, managers across the construction program and the supervisors. This dialogue did not radically change what was done by supervisors in their daily work, starting with a "Daily Activity Briefing" to their teams, which sponsored the first questions of directors during their tours of the sites: "Have you had a briefing this morning? What did it cover? What do you think? What's it like working here?" and so on. The supervisors remained at the heart of permits to work, risk assessments, and method statements, developing and briefing them out at the start of each phase of work. Supervisors did become more involved in incident reporting and investigations, some attending sessions on how to explore incidents to search out systemic failings that could be corrected to prevent a recurrence. During three years of major construction, the supervisors did what supervisors do in the day-to-day management of their teams, but they did it in a dynamic way so that they really did function as the major component of site works management. It is difficult to be more precise, although the research on leadership on the program already cited does seek to do so. Essentially, supervisors ceased to simply follow and transmit orders. They became much more active in leadership and management of the program and this had a measurable effect on the identification, elimination, and management of risks.

8.7 EMPOWERMENT, PEDESTALS, AND PRIDE

Even where supervisors have been identified in the past as key players in the stories of health and safety on site, the approach has been instrumental – developing skills to oversee the work day-to-day and to intervene to achieve compliance with rules. The program for Black Hats on the London 2012 construction works was broader. The ODA and SHELT actively sought to encourage supervisors to speak up, so that their teams would stay safe. There was a conscious effort to empower them, to require their managers to promise to respond to their communications through genuinely listening to their concerns and suggestions. This was matched by continually telling the supervisors that they were respected, were important, were the key leaders on site – this process effectively put supervisors "on a pedestal" and was a mechanism that put them under a great deal of "soft" pressure to live up to their role – to deliver on their responsibilities. As they did so and the health and safety performance on London 2012 projects began to set new standards for UK construction, there was a palpable sense of pride in doing their jobs well, which was reinforced by the reward and recognition schemes operating on every project and across the program. The desire to sustain the good record, to not be the individual or the team that sullied that record, became a very powerful driver for good performance. It also changed the perception from health and safety being about following rules to it being an expression and proof that each job was being done in the right way – that is why Lord Dannatt in his lecture was able to cite it as the underpinning to completing the works on time, within budget, and to a high standard of quality.

8.8 OUTCOMES

Construction site work is never a controlled experiment. The London 2012 teams employed every mechanism that they could identify, either directly or in pilots and trials, to seek to prevent harm and enhance wellbeing. These included: development of the "Health and Safety Standard" and weaving it into the procurement process and the contracts with the supply chain; driving designers' compliance with the requirements for safe and healthy by design, with a risk management design coordinator (Scopes 2009); setting up and supporting a leadership team across the whole supply chain; establishing an on-site occupational health service at a time when few construction companies had any significant resources to deploy to manage health risks (Tyers and Hicks 2012); on-line reporting of incidents; and a leadership team overview of trends and incident analysis and response (Shiplee et al. 2011). The wide range of initiatives and programs of action is indicated in a series of papers which remain accessible as a national archive (UK's National Archive 2012; 2016; 2018). Clearly, the outcomes cannot be attributable to any one aspect of the health and safety efforts, including the development of Black Hats as leaders.

The results of all the efforts was the first Olympic construction program without a work-related fatality. There were two deaths on site, in both cases individual workers succumbed to health conditions, a stroke and a heart attack. The overall reportable accident rate for the whole period was close to 0.2, that is a reportable accident for each half million hours worked, and for the final two years of the program and the two years of Olympic Village and Olympic Park conversion, employing the same leadership and management approach, the rate was below 0.1. In addition to the gross numbers, the program generated a sense of real pride amongst the site workers, as evidenced by surveys, and created across the UK construction industry a palpable confidence that zero harm could be more than a slogan and was both realistic and achievable.

The development of supervisors as leaders has become a standard part of many large projects and major construction companies' approach to managing risk and addressing wellbeing. From as far afield as the Canterbury Rebuild and its "Safety Charter" launched after the devastating earthquakes in New Zealand in 2011 (Stats NZ 2018; Canterbury Safety Charter 2019) to projects elsewhere in London, the methods for developing supervisor leadership have been recognized as having value and to be emulated. For example, a major university construction program established in the years following the London 2012 works commissioned a training program that focused on elements drawn from the previous experience and covered:

- Recognizing the role good supervision plays in developing a positive culture
- Identifying the triggers for and challenging inappropriate behavior and attitudes
- Developing behaviors to enhance and support a positive culture
- Demonstrating communication skills to deliver effective and meaningful daily task briefings and toolbox talks
- Mastering reward and recognition

This has been taken on, as an example, by an international materials company, with its "First Line Leadership training program". Its UK arm has centered on developing interpersonal and communication skills by coaching the use of body language and voice to improve interactions with their teams, and also with workers not in their direct line of management such as contractors. The soft skills that previously were not included in traditional safety and health management courses are now mainstream, a direct legacy of the London 2012 Black Hat program. On construction programs in areas such as the Middle East, Far East, and Africa, the same approach, recognizing that first-line supervisors often have social leadership and translation roles for their teams, can also be applied. That is because this approach is based on the reality that this layer of workers, the supervisors who are at the interface with formal line management, can be the most influential as to how the work is done and what information is passed upwards to influence strategic decision-making. The success of the Black Hat program may well prove to be more than a localized and specific result, but rather a pathfinder for a more general re-evaluation of where leadership truly lies and how it may be developed. Vision Zero, the campaign launched by the International Social Security Association (ISSA) at the World Congress for Safety and Health in 2017 in Singapore, envisages a world without accidents and ill health caused by work. While senior management commitment across organizations is a key initiating factor in achieving this, it may only be realized when the whole workforce is engaged – and the London 2012 Black Hat program has demonstrated that supervisors are key to doing so.

REFERENCES

Build, U. K. 2019. *Leading the Construction Industry.* https://builduk.org/ (accessed on December 14, 2019).

Canterbury Safety Charter. 2019. https://safetycharter.org.nz/.

CCS – Considerate Constructors Scheme. Ware, UK: CCS. https://www.ccscheme.org.uk/ (accessed on December 14, 2019).

Constructing Excellence. 2015. *Respect for People.* Garston Watford, UK: Building Research Establishment Ltd. http://constructingexcellence.org.uk/category/respect-for-people/ (accessed on December 14, 2019).

Courtney, W. H. 2000. Constructions of Masculinity and Their Influence on Men's Well-Being: A Theory of Gender and Health. *Social Science and Medicine* 50(10), 1385–1401. https://www.sciencedirect.com/science/article/abs/pii/S0277953699003901 (accessed on December 14, 2019).

CPS – United States Department of Labor. 2010. *Labor Force Statistics from the Current Population Survey.* Washington, DC: Bureau of Labor Statistics. https://www.bls.gov/cps/.

CPWR – The Center for Construction Research and Training. 2013. *The Construction Chart Book. The U.S. Construction Industry and Its Workers*, 5th ed. Silver Spring: CPWR. https://www.cpwr.com/sites/default/files/publications/5th%20Edition%20Chart%20Book%20Final.pdf (accessed on December 14, 2019).

Egan, J. 2014. *Rethinking Construction.* London: HMSO Department of Trade and Industry. http://constructingexcellence.org.uk/wp-content/uploads/2014/10/rethinking_construction_report.pdf and https://www.designingbuildings.co.uk/wiki/Egan_Report_Rethinking_Construction (accessed on December 14, 2019).

Hinzie, J., and P. Raboud. 1988. Safety on Large Building Construction Projects. *Journal of Construction Engineering and Management* 114(2). doi:10.1061/(ASCE)0733-9364(1988)114:2(286).

HSE – Health and Safety Executive. n.d. Health and Safety Statistics. http://www.hse.gov.uk/statistics/ (accessed on December 14, 2019).

IES – Institute for Employment Studies. 2013. *Leadership and Worker Involvement on the Olympic Park.* London: Olympic Delivery Authority (ODA). https://webarchive.nationalarchives.gov.uk/20130403014821/http://learninglegacy.independent.gov.uk/public

ations/leadership-and-worker-involvement-on-the-olympic-park.php   (accessed   on
    December 14, 2019).
IES – Institute for Employment Studies. 2013a. *Learning Legacy: Lessons Learned from
    the London 2012 Games Construction Project.* London: Olympic Delivery Authority
    (ODA).   https://webarchive.nationalarchives.gov.uk/20130403014831/http://learningle
    gacy.independent.gov.uk/publications/occupational-hygiene-at-the-olympic-park-an
    d-olympic-and.php (accessed on December 14, 2019).
IES – Institute for Employment Studies. 2016. *About Learning Legacy.* London: Olympic
    Delivery Authority (ODA). https://webarchive.nationalarchives.gov.uk/201610031144
    01/http://learninglegacy.independent.gov.uk/about/ (accessed on December 14, 2019).
ISSA – International Safety Security Association. Vision Zero Campaign, Geneva: ISSA.
    http://visionzero.global/ (accessed on December 14, 2019).
Latham, M. 1994. *Constructing the Team.* London: HMSO Publications Centre. http://con
    structingexcellence.org.uk/wp-content/uploads/2014/10/Constructing-the-team-The-
    Latham-Report.pdf (accessed on December 14, 2019).
ODA – Olympic Delivery Authority. 2012. ODA Health, Safety and Environment Standard.
    https://webarchive.nationalarchives.gov.uk/20120403121557/http://www.london2012
    .com/publications/oda-health-safety-and-environment-standard.php   (accessed   on
    December 14, 2019).
Prescott, J. 2001. Six months to improve safety. Speech at the Health and Safety Commission
    construction safety summit. London.
Office for National Statistics. 2010. *EMP13: Employment by Industry (Labour Force
    Survey).* Construction 2005-2012 as example for the period of London 2012. Newport /
    Titchfield/London. https://www.ons.gov.uk/employmentandlabourmarket/peopleinwor
    k/employmentandemployeetypes/datasets/employmentbyindustryemp13 (accessed on
    December 14, 2019).
Scopes, J. P. 2009. London 2012: A New Approach to CDM Coordination. *Proceedings
    of the Institution of Civil Engineers – Civil Engineering* 162(2), 76–86. doi:10.1680/
    cien.2009.162.2.76
Seely, A. 2018. *Self-Employment in the Construction Industry.* Briefing paper no 196.
    London: House of Commons Library. https://www.parliament.uk/commons-library.
Serpell, A., and X. Ferrada. 2007. A Competency-Based Model for Construction Supervisors
    in Developing Countries. *Personnel Review* 36(4), 585–602.
Shiplee, H., L. Waterman, K. Furniss, R. Seal, and J. Jones. 2011. Delivering London 2012:
    Health and Safety. *Proceedings of the Institution of Civil Engineers – Civil Engineering*
    164(6), 46–54.
Smith, D., and P. Chamberlain. 2015. *Blacklisted. The Secret War between Big Business and
    Union Activists.* Oxford: New Internationalist Publications Ltd.
Stats NZ (2018). Canterbury: the rebuild by the numbers. Retrieved from www.stats.govt.nz.
Tyers, C., and B. Hicks. 2012. *Occupational Health Provision on the Olympic Park and
    Athletes' Village.* Research Report RR 921. Bootle: Health and Safety Executive. www.
    hse.gov.uk/research/rrhtm/rr921.htm (accessed on December 14, 2019).
Tyers, C., S. Speckesser, B. Hicks, K. Baxter, M. Gilbert, and E. Ball. 2012. *Occupational
    Hygiene at the Olympic Park and Athletes' Village: Can Workplace Health Management
    be Cost Effective?* Report 497. Brighton: Institute for Employment Studies. https://ww
    w.employment-studies.co.uk/resource/occupational-hygiene-olympic-park-and-athlet
    es-village (accessed on December 14, 2019).
UK's National Archive. 2012. https://webarchive.nationalarchives.gov.uk/20120403073304/
    https://www.london2012.com/ (archived on April 3, 2012; accessed on December 14,
    2019).
UK's National Archive. 2016. *About Learning Legacy.* Kew, Richmond, Surrey, UK. https://
    webarchive.nationalarchives.gov.uk/20161003114401/http://learninglegacy.independen
    t.gov.uk/about/ (accessed on December 14, 2019).

UK's National Archive. 2018. Common Standard for the Supervisors Training Course. http://learninglegacy.independent.gov.uk/publications/common-standard-for-the-supervisors-training-course.php.

Waterman, L. 2014. *The London 2012 Construction Project – Beyond Expectations. Hong Kong Construction Industry Council: Construction Safety Week 2014.* Hong Kong: DEVB & CIC. https://www.cic.hk/cic_data/pdf/about_cic/news_and_update/past_event/chi/Lawrence%20WATERMAN.pdf (accessed on December 14, 2019).

9 Workers in a Virtual Work Environment

An Immersive Safety Learning Experience

Bonnie Yau, Toran Law, and Steve Tsang

CONTENTS

9.1 INTRODUCTION

In order to enhance training quality and facilitate more effective training outcomes, i.e., to modify or reinforce the attitude of trainees towards safe and healthy working behavior, the Occupational Safety and Health Council (OSHC), a statutory body for promoting safety and health at work and sustaining the valuable workforce of Hong Kong, introduced a number of experiential learning elements into its training courses several decades ago. To explore a better and more effective method for safety and health knowledge transfer and building on its experiences of delivering experiential learning, the OSHC established the "OSH Immersive Experience Hall" in its OSH Academy in 2018. The study here reports a practical case of adopting immersive virtual reality (VR) technology to provide safety training in working at height for the construction industry. The design and implementation of the immersive system as well as its limitations are discussed.

9.2 SAFETY AND HEALTH TRAINING

Provision of effective safety and health training to workers plays a crucial role for safety knowledge transfer to ensure work-safe competences and hence a preventive culture.

9.2.1 COMMON SAFETY AND HEALTH TRAINING METHODS

There are several ways to transfer safety and health knowledge to workers in training. The Health and Safety Executive of the UK suggested a number of safety and health training methods in its "Health and Safety Training: A Brief Guide" (HSE 2012) which includes the following:

- Giving information or instruction
- Training in the classroom
- Coaching or on-the-job training
- Open and distance learning
- Computer-based interactive learning

The first three methods are traditional safety and health training approaches that have been used for decades. Due to the rapid development of information technology and digitization in recent years, open and distance learning of safety and health knowledge is possible. It enables workers to receive safety and health knowledge via the internet from their computers or various handheld devices. Computer-based interactive learning, for example, transfer of safety and health knowledge by computer animation or interactive games, provides vivid impressions to keep workers attentive and engaged in the training process.

Burke et al. (2006) have conducted a meta-analysis study to compare the relative effectiveness of safety and health training methods among 95 studies conducted between 1971 and 2003 in 15 countries. They categorized the safety and health training methods into three levels according to their engaging effects. Lectures, films, and video-based training were categorized as the least engaging; computer-based instruction and feedback techniques were moderately engaging; while behavioral modeling, simulation, and hands-on training were most engaging. They found that when occupational safety and health training methods involved a high level of engagement or active participation by the workers, greater knowledge acquisition and significant reductions in negative outcomes such as fewer injuries and accidents, as well less illness, were observed. A later meta-analysis by Burke et al. (2011) indicates that safety knowledge acquisition and safety performance was affected by the interaction between the levels of engagement of safety training and the severity of the hazardous event/exposure. That is, when the severity of the hazardous event/exposure was high, highly engaging safety training enabled a more effective development of safety knowledge and performance than did the less engaging one. Such findings highlight the necessity of the design and implementation of participatory/interactive approaches, to safety and health training, so as to enable an active involvement of workers to facilitate the transfer of safety and health knowledge for the achievement of a better safety performance in the workplace.

9.2.2 EXPERIENTIAL LEARNING

Engaging methods of safety and health training may be well explained by Experiential Learning Theory (ELT). According to ELT, learning is defined as "the process whereby knowledge is created through the transformation of experience. Knowledge results from the combination of grasping and transforming experience" (Kolb 1984, 41). ELT is a four-stage learning process as illustrated by the Kolb's experiential learning cycle (Kolb 1984; see Figure 9.1).

The four-stage learning cycle starts with "Concrete Experience", where learners are exposed to a new experience or situation. In the second stage of "Reflective Observation", learners review what they have just experienced, and reflect on any inconsistencies between the new experience and their prevailing understanding. In the third stage, "Abstract Conceptualization", learners try to conceptualize what they have learned and form a "theory" or "model" for subsequent hypothesis testing.

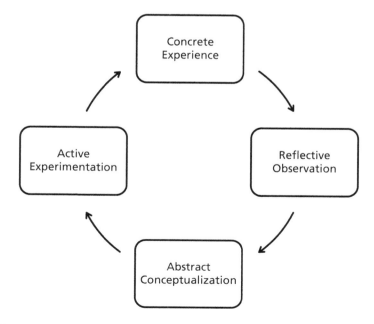

FIGURE 9.1 Kolb's experiential learning cycle (Kolb 1984).

During the final stage of "Active Experimentation", learners try to put into practice what they have learned (hypothesis testing) and form new experiences. From this perspective, learning is an integrated process of the four stages where all stages are mutually supportive and feeding into the next. Effective learning could only be realized when all four stages have been executed and when the logical sequence of the learning cycle is followed regardless of the starting stage.

Researchers found that experiential learning is proven to be an effective way of enhancing knowledge transfer and developing positive attitudes of trainees (Pugsley and Clayton 2003; Maslen 2014; Angelopoulou et al. 2015). For example, Pugsley and Clayton (2003) compared the attitudes between two groups of nursing students. They found that students who were in the experiential course exhibited significantly more positive attitudes towards nursing research than those in traditional classroom courses.

9.2.3 APPLYING EXPERIENTIAL LEARNING IN SAFETY AND HEALTH TRAINING

Under conventional safety and health training methods, such as classroom training, safety practitioners would talk about safety and health knowledge and the factual consequences of fatal accidents. These training methods have been criticized of providing low level of engagement in safety training and cannot ensure effective safety knowledge transfer for desired safety performance behaviors (Burke et al. 2006). Applying experiential safety and health training can possibly fill this gap. Imagine if workers have a chance to experience the negative, dire consequence of risky behavior: despite that, they need not go through the real hazard, they will learn from what they have personally experienced, which results in activation of the experiential learning cycle (Kolb 1984). Besides, such near-real experience will allow them to experience the severe consequence of the risky behavior, which may reduce their risk-taking propensity, ultimately increasing their safety motivation. This effective change in safety knowledge and safety motivation might help them to be more prone to implementing safety measures when they are back to their workplaces. Safety practitioners in some countries already adopt experiential learning in their safety and health training program for workers. For example, the Safety Education Training Center of Sumitomo Heavy Industries Marine & Engineering Co Ltd., located in Yokosuka in Japan, has a number of experiential learning facilities. Trainees will be asked to experience the unsafe situations using these facilities, such as staying on a vibrating ladder and walking on a gangway with trip and slip hazards under a controlled environment where safety measures are implemented to protect trainees during their experiential learning, for example, trainees are required to grasp the hand-rails when they are walking on the gangway with trip and slip hazards, so that they will not fall but at the same time can experience the dangers of slipping and tripping.

9.2.4 LIMITATION OF CONVENTIONAL EXPERIENTIAL LEARNING FACILITIES

Now we know that experiential learning is an effective tool for transferring safety and health knowledge. The next question is how we can maximize the effectiveness of such an approach, making it even more powerful, to achieve better safety and health training outcomes. Here, some fundamental changes to the existing mode of safety learning as experience may be called for. Conventional experiential learning facilities have their limitations – they cannot expose trainees to extreme and extraordinary dangerous situations and, at the same time, ensure trainees' safety for various training purposes. For example, pilot training in the real world is usually costly from financial and environmental perspectives, not to mention the safety concern of exposing pilots under demanding and emergency situations for flight training in reality. Therefore, flight simulators were developed for pilot training. Similarly, for construction safety training, given the hazardous nature of construction sites, onsite training is dangerous and the cost of failure is tremendous, making training through experience of failure/accident almost impossible (Sacks et al. 2013): For example, to build a substandard scaffold for workers to experience the consequence of working without safety harness on a collapsing scaffold could itself create real hazards. If training through experiential learning in real life is unfeasible and unsafe, it is important to think of an alternative method that can physically simulate the real-world environment for various training tasks to allow trainees to learn, interact, and experience in a safe and controlled environment.

9.2.5 THE ADVENT OF VIRTUAL ENVIRONMENT-BASED TRAINING

Thanks to the rapid development of internet and communication technology (ICT), data exchange between physical and cyber systems can be realized in many applications. Among those, virtual environment-based trainings have drawn a lot of attention and it has been proven to be able to break the limitations of existing experiential learning facilities and evolve safety learning (Gupta et al. 2008; Christou 2010). A virtual environment can be defined as a computer-generated environment for the simulation of the real world and it can either be semi-immersive, i.e., a virtual environment with a two-dimensional (2D) interface, or full immersive, i.e., a virtual environment with a three-dimensional (3D) interface (Ong and Nee 2004). The unique characteristics of a virtual environment offers a wide variety of possibilities and advantages for training applications. First of all, different from text-based or video-based instructions, which are usually passive and non-interactive in nature, a virtual environment allows extensive user interactions that can greatly enhance retention of spatial information of the users. In addition, a virtual environment enables trainees to have a 3D immersive experience, which can facilitate a stereographic projection of spatial relationships that can hardly be achieved by 2D displays. Further, a virtual environment supports multi-sensory inputs and outputs that enable users to have different modality interactions (e.g., visual, auditory, and/or tactile).

9.2.6 VIRTUAL REALITY TECHNOLOGY AS A SOLUTION

VR refers to a specific form of a virtual environment providing the user with a feeling of presence. The feeling of presence requires users to immerse in the virtual environment by feeding them with a continuous stream of realistic stimuli (e.g., visual, auditory, and/or tactile stimuli) and enabling them to easily interact with the virtual objects (Gupta et al. 2008). The immersive nature of VR has enabled its wide application to virtual environment-based training to facilitate the cognitive and spatial learning of the user. These applications encompass a wide spectrum of areas including medical surgery (Gallagher et al. 1999; Gurusamy et al. 2009; Thomsen et al. 2017), manufacturing assembly (Seth et al. 2011; Gavish et al. 2015; Matsas et al. 2017), military force (Moshell 1993; Lele 2013), and safety training in different industries (Filigenzi et al. 2000, Sacks et al. 2013).

By applying VR technology for experiential learning, workers are able to completely immerse themselves in a virtual environment which mimics a realistic work environment for learning and training under different preset scenarios. These training scenarios can be a variety of daily tasks to be performed by the workers. Therefore, it can allow the worker to master the task before actually performing it. It is particular advantageous to adopt the virtual environment-based training for some critical tasks that a failure or mistake made during the actual work could cause catastrophic consequence (e.g., loss of life, personal injury, and/or property damage). Besides, the realistic and immersive virtual environments can allow the worker to experience a serious work accident under a safe circumstance. Such life-like accident experience is expected to be able to provide the deep psychological drive for participants to identify hazards on the job due to realistic training. Also, VR offers a time efficient training method to the industry as unlimited repetitions of training can be easily achieved. For example, critical working procedures can be practiced repeatedly in a virtual environment safely and effectively. Another advantage of using VR training is that it allows trainees to know the result of their wrongdoing instantly. The trainer can allow trainees to act freely in the virtual environment without stopping them even when they are engaged in some dangerous acts. The VR system presents the trainee with the consequences of what he or she has done individually. This function provides trainees the best way to learn to avoid making mistakes and strongly motivates them to follow safety procedures and practice safe work behavior in their workplace in the future. All these capabilities demonstrate that VR is a useful tool for safety learning as experience in various industries (Gupta et al. 2008; Christou 2010; Pantelidis 2010).

9.2.7 KEYS TO APPLYING VR TECHNOLOGY

VR technology is not a "magic wand" that can improve the quality of training dramatically overnight. It is, like other aids and technologies, a tool only. What is crucial in making VR a success is a good combination of the VR technology and a high quality safety and health training program at the start, which provides state-of-the art knowledge, including interactive learning and learning by doing, and takes care for a fit between educational activities and devices and the learning objectives (Donabedian 1966; Swuste and van Dijk 2019). If it is therefore justified to apply VR technology in safety and health training, we should consider which VR system should be selected. Recently, two types of systems are commonly applied in different VR training programs: Head-Mounted Display (HMD) and Cave Automatic Virtual Environment (CAVE). The former displays the simulated environment to the user by a head-mounted device worn by trainees; while the latter projects the simulated environment in a cube of projection screens and the trainee in the cube can view it by wearing a pair of 3D glasses. Effects of various types of display technologies for VR training have been extensively studied over the years. For instances, Kim et al. (2014) compared the effects of different virtual environment technologies of a standard monitor, HMD, and CAVE on emotional arousal and task performance. Their results showed that using HMD and CAVE displays could lead to higher emotional arousal than a monitor display. Also, among the display systems, CAVE induced the highest sense of presence, while HMD resulted in the highest amount of simulator sickness. In fact, both systems have their own advantages and limitations. First, HMD is far cheaper and requires less physical space. Yet, the display device worn by the user is bulkier and the user's vision is completely isolated from the surroundings and his or her own physical presence. Discomfort, such as nausea, is also frequently reported with the use of HMD (Porcino et al. 2017). On the other hand, CAVE is more expensive and requires more space to set up. The 3D glasses worn by the user are light-weight and offer a larger field of view (Mestre 2017). Users can move freely inside the display cube and have a vivid interaction with his/her body parts (as the user can see them) during training. In fact, they would be able to sense their presence in the make-believe environment and the impressions are generally more lasting. If we want to let users experience an accident in a virtual work environment during safety and health training, CAVE seems to be a preferable option despite being more costly.

9.3 METHOD

9.3.1 PARTICIPANTS

One hundred and six participants from the public safety courses and events orga-nized by OSHC were invited to participate in the VR immersive training. As reported in previous research, discomfort symptoms, such as nausea, sickness, and headaches may result from immersive VR environments. Prior screening was there-fore conducted to ensure the students were medically and physically fit to participate in immersive VR training.

9.3.2 APPARATUS AND DESIGN

There are many types of CAVE systems available in the global market. In Hong Kong, a commonly used type is the imseCAVE developed by the Department of Industrial and Manufacturing Systems Engineering of the Faculty of Engineering, University of Hong Kong (HKU 2015). The imseCAVE has been successfully applied for the creation of different virtual environments to provide immersive training experiences to various users (Lau et al. 2007; Yuen et al. 2010). It is a fully immersive and inter-active visualization system that provides extremely vivid stereoscopic views of an unlimited virtual world within a small room of a few square meters. The system offers a sense of reality and flexibility that expands human perception of reality and virtuality. Interactive devices such as data gloves, trackers, joysticks, and motion platforms can be readily connected to the imseCAVE for better immersive effects and interactivity of the virtual environment, which makes it suitable for creating the virtual work environment and simulating the work accidents used in the VR experiential safety and health training. The major part of it is a cube-like structure consisting of four projection screens, i.e., three sides (front, left, right) and bottom, a metal truss for the mounting of four high resolution 3D projectors, 12 high precision motion trackers, and a surround sound system. This structure is connected to a high-speed computing and imaging processing server, so that 3D images and sound effects can be projected inside it to create the near-real virtual environment. Motion trackers are used for tracking the user's motion to produce interactive visual effects; 3D glasses and interaction devices are also essential, where the former is to make the users' view more vivid; and the latter is for users to interact and perform tasks. Here, the imseCAVE system was used to create an immersive virtual environment for safety training.

9.3.3 TRAINING SCENARIOS

Working at height is a high-risk operation that causes a considerable number of fatal accidents in Hong Kong, particularly in the construction industry. In 2018, there were 11 fatal industrial accidents due to falling from height in the construction industry, accounting for 78.6% of the total fatal industrial accidents in the industry (LD 2019). Though much effort has been made by different stakeholders, including the Labor Department of the Hong Kong Government and OSHC to promote safe work at height, such as a subsidy scheme for small and medium enterprises to purchase different safety devices (e.g., ladder platform) and personal protective equipment (PPE) (e.g., transportable temporary anchorage devices), fatal cases of falling from height still happened. It is obvious that a lack of a safety device and PPEs are not the main causes of accidents but rather more on the awareness of the importance and proper use of these devices. Hence, in the mandatory basic safety training for construction workers, OSHC offers every trainee the chance to wear a full body harness. Yet, while training could instruct on how and when to use PPEs correctly, the consequence of failing to use safety harness was only delivered through traditional methods such as lectures, video, storytelling, and so on. With the imseCAVE system, the accident scene could be vividly captured and simulated. VR could be an apt way for a trainee to personally experience the danger of working at height. Hence, working at height safety was selected as the first scenario to be developed. It is not only about an unsafe work at a height situation but also about a work setting that is familiar to the trainees.

After confirmation of the scenario theme, OSHC and HKU worked closely to develop the storyboard. A typical working scene reproduced by the VR environment shows renovation work on the external wall of a high-rise residential flat with a truss-out bamboo scaffold as shown in Figure 9.2. Site visits to renovation works were arranged to get all parties, especially the image rendering designers and system developers of HKU, familiarized with the actual working environment and the relevant working procedures. After gathering all the information, a simulated truss-out bamboo scaffold work environment was developed as shown in Figure 9.3. During the training, the scaffold would suddenly fall apart as one of the supporting metal brackets at the bottom detached from the external wall due to improper installation

FIGURE 9.2 A truss-out bamboo scaffold used in Hong Kong for renovation works at external walls of buildings.

FIGURE 9.3 A near-real simulated truss-out bamboo scaffold work environment with a supervisor giving instructions.

to allow the trainee to experience falling from height with his supervisor as illustrated by Figure 9.4. Aided by the visual and sound effect generated, a near-real sensation of falling could be experienced, and the experience on the consequences of unsafe behavior (i.e., not using the fall arresting system) can have a long lasting and impactful effect on the memory of the trainee. Safety practitioners were involved to give comments at different preparatory phases to ensure not only the effects were real but also the interaction within the VR system was as authentic as possible.

FIGURE 9.4 Participants experience falling from height together with a supervisor.

Besides experiencing a sad ending, participants may wish to know what would have happened if they had used the fall arresting system. Therefore, the scenario also allows the participants to experience the alternative ending by wearing a real safety harness that can prevent them from falling from height. No matter which ending the trainee experienced, at the end of the scenario, a 3D video on proper safety measures that should be implemented is presented, including proper erection of the truss-out scaffold, use of the fall arresting system, and so on.

9.3.4 PROCEDURE

Before the training session, the participant needed to wear a pair of 3D glasses and the interactive devices for immersion into the simulated work environment. After that, an instructional video was displayed to the participant introducing the simulated work environment and the tasks they were going to conduct therein. The participant then followed the instructions provided by the system to conduct the task like the practitioner would be doing in the actual work environment. Specifically, first of all, the participant was asked to stand on the virtual scaffold and conduct a daily job task. In the course of working, the system created a work accident scenario where the participant, without using any fall arresting system, would suddenly fall from the scaffold, allowing the participant to experience the serious consequence of non-compliance with safe practices and risky behavior at work. Following that, the system showed the safety measures that should have been implemented in order to prevent the accident that had just happened. Before leaving the Hall, the participant was asked to complete a questionnaire to provide his/her feedback about the experience of using the imseCAVE system and his/her perceived usefulness of the immersive training for improvement of their safe behaviors at work.

9.3.5 PARTICIPANTS' FEEDBACK EVALUATION

Participants of the OSH Immersive Experience Hall were requested to complete an evaluation questionnaire to provide their feedback on their experiential learning in the Hall using an iPad. The evaluation questionnaire contains seven questions about the training experience of the users such as "The Hall increased my interest for learning working at height safety knowledge" and the effectiveness of enhancing their working at height safety knowledge such as "The Hall enhanced my knowledge on working at height safety" and "After using the Hall, I now care more about working at height safety". Responses were rated on a five-point Likert scale where 1 represents "Strongly Disagreed" and 5 represents "Strongly Agreed".

9.4 RESULTS AND DISCUSSION

9.4.1 FEEDBACK FROM THE PARTICIPANTS

One hundred and six participants took part in the immersive VR training. Their feedback is summarized in Table 9.1. As shown in the table, the mean score of all the evaluated questions were greater than 4.4 out of 5, showing that most participants agreed or strongly agreed that the OSH Immersive Experience Hall (1) gave them a near-real experience; (2) enhanced their knowledge on working at height safety; (3) enhanced their interest for learning working at height safety knowledge; and (4) offered a better learning experience than classroom lecturing. Most of the users also agree that they would (5) care more about safety when working at height; and (6) be more willing to remind others about the importance of working at height safety after the experience. The results showed that overall, the participants liked to use the Hall to learn safety and health knowledge with a mean score of 4.58. Besides, Spearman's rank-order correlation was run to determine the relationship between each of the questions. All the questions had strong and positive correlations between each other, and these correlations were statistically significant ($ps \leq 0.01$).

9.4.2 EFFECTIVENESS OF IMMERSIVE VR EXPERIENTIAL LEARNING

The immersive nature of VR experiential learning makes it a promising method for safety training. A critical review by Li et al. (2018) found that virtual reality and augmented reality (AR) had been widely used in construction safety, and those applications can be mainly categorized as hazard recognition and identification, safety training, and education, as well as safety inspection and instruction. Our findings in using the imseCAVE system are in line with similar results reported by Sacks et al. (2013) and Perlman et al. (2014). For example, Sacks et al. (2013) evaluated the effectiveness of construction safety training using immersive VR technology, and they found that VR training was more effective than traditional classroom training methods as trainees would more readily retain their attention and concentration with VR learning. They reported that VR training could arouse the alertness of the trainees to a larger extent and for a longer period of time compared with the normal training in which concentration of the trainees was able to be maintained only in

TABLE 9.1

Participants' Feedbacks on Experiential Learning in OSH Immersive Experience Hall (Means, Standard Deviations, and Correlations)

	Mean	SD	1	2	3	4	5	6
OSH Immersive Experience Hall gave me a near-real experience.	4.49	0.68						
The Hall enhanced my knowledge on working at height safety.	4.53	0.59	0.615**					
The Hall increased my interest in learning working at height safety knowledge.	4.45	0.71	0.629**	0.545**				
The Hall offers a better learning experience than classroom lecturing.	4.58	0.63	0.599**	0.569**	0.651**			
After using the Hall, I now care more about working at height safety.	4.59	0.64	0.612**	0.524**	0.699**	0.767**		
After using the Hall, I am now more willing to remind others about the importance of working at height safety.	4.60	0.61	0.567**	0.492**	0.643**	0.666**	0.669**	
I like to use the Hall to learn safety and health knowledge.	4.58	0.62	0.605**	0.577**	0.547**	0.617**	0.592**	0.648**

$**p \leq 0.01$.

the first hour. As indicated in the evaluation feedback, most participants agreed that compared with conventional classroom lecturing, the immersive training was able to enhance their learning experience, so that their interest for learning working at height safety knowledge was greatly improved.

In fact, applying VR technology such as the imseCAVE system in experiential learning for safety and health training has a number of advantages. It offers trainees a simulated work environment which provides trainees a feeling of presence by feeding them with realistic stimuli, and supports multi-sensory inputs and outputs that enable trainees to have multi-modality interactions with the virtual environment for an enhanced training experience (Freina and Ott 2015). Such near-real and interactive environments not only arouse the interest of trainees for learning safety and health knowledge, but also gives them a stronger impression and better understanding of the training content so as to maximize learning effectiveness (Yuen et al. 2010). Besides, work accidents can be simulated by the VR technology and trainees can have a life-like accident experience without really harming themselves. This experience would be difficult to create, if possible, for other modes of experiential learning. As a result, trainees are allowed to assess a hazardous situation, make a decision on a series of actions, execute an action, and immediately observe the consequence of their action. Even if a wrong decision is made, no harmful effects result. This learning experience has the potential to result in cognitive information processing and is conducive to a more long-lasting impact on the trainees for the safety and health knowledge they have received (Lucas et al. 2008). The evaluation feedback obtained here also indicated that almost all participants agreed that they were more concerned with height safety after experiencing immersive VR training, such an increment in their awareness would be expected to eventually lead to higher motivation to comply with the safe practices for working at height activities at work.

Moreover, trainers can create different training scenarios according to the trainees and the training needs, providing plenty of opportunities on safety and health training themes within a limited space. This is especially beneficial for trainers when training space is limited or lacking. Jeelani et al. (2016) found that applying personalized hazard recognition training in the construction industry, where training strategies were developed based on the knowledge gaps and learning needs of particular workers, was able to increase the average number of recognized hazards from 42% to 77%. However, one of the limitations reported by the authors was the use of 2D static images, which were unable to capture the true dynamic nature of construction operations. In order to overcome this limitation, they later successfully developed an immersive virtual environment that was able to offer a higher degree of realism of construction operations (Jeelani et al. 2017), indicating the high flexibility of customization or personalization of the training scenarios using VR technology. Here, the working at height scenario was created for the provision of immersive experiential learning to the participants. The evaluation results showed that the majority of the participants agreed that the imseCAVE system gave them a near-real experience for working at height safety. In the future, more construction safety scenarios (e.g., lifting operations) should be created for various trades of workers in the construction industry.

9.4.3 STUDY LIMITATIONS AND CHALLENGES AHEAD

In this study, most participants provided very positive feedback of their immersive learning experience with the imseCAVE system. However, it should be noted that the evaluation method here may result in the bias of social preferable answers and may not be able to truly reflect the effectiveness of the training. Therefore, in the future, other indicators shall be used, such as the number of working at height accidents/incidents after the training, and as far as there may be a problem with the reliability of this indicator, the impact of the training on the accident process itself and the quality of the measures to prevent accidents (Swuste and van Dijk 2019).

In fact, although there are many advantages of applying VR technology for various safety and health training purposes, challenges do exist. One of the key challenges is to identify the right scenarios for further development. These future scenarios may take into account additional aspects like the impact of a no-blame-culture on the behavior of workers (e.g., its impact on the motivation to report near-misses and accidents) and the influence of the working environment (e.g., work organization and leadership).

Another major challenge is resource availability. Applying VR technology in safety and health training requires considerable investment in hardware, software, and programming. The initial resource requirement is much higher than conventional training modes like classroom lecturing. To maximize the return on investment of adopting VR technology, organizations can collaborate with other stakeholder parties in developing training scenarios that are of mutual interest. They may also exchange developed scenarios providing that the costs involved in the calibration of the system are within a reasonable range. The sharing of scenarios allows trainees of both organizations to experience and practice situations, resulting in a better use of resources in terms of cost and time.

In addition, making a simulated work environment near-real is also a big challenge. The higher the authenticity of the simulations, the better the impact of the relevant safety and health training. This requires meticulous information gathering during the development of the scenarios, including site visits, the study of relevant accidents, and interviews with practitioners. Accurate information gathering at this preparation stage is crucial to the overall success of experiential training. Moreover, where the budget allows, the system should also add on extra dimensions, such as a motion platform or wind blower to increase the sense of reality. Sound effects should also be perfected to create the near-real environment. Although the existing basic settings of the imseCAVE system in the OSH Immersive Experience Hall are able to produce near-real accident experience to trainees, the degree of reality can be enhanced by further modifications of the system. In the next phase of development, a motion platform will be installed under the floor screen, which will interact with the scenario, to give trainees an extra dimension of experience i.e., a 4D experience. For example, the platform will vibrate and move before the scaffold collapses; and a fast downward motion and vibration at the time the trainee "falls down" will make the experience even more intensive and bring the whole experience to the next level. We believe through these modifications, trainees will have a stronger impression and achieve even better training outcomes.

9.5 CONCLUSION

Experiential learning provides an engaging method for education and training that enables effective knowledge transfer. In this chapter, a case study of integrating virtual reality technology with experiential learning for provision of safety training using the imseCAVE system was presented. The feedback from the users of the system was positive and all agreed that the virtual environment-based training increased their awareness of dangers at work. The application of VR technology, in particular, the imseCAVE system introduced here, can maximize the effectiveness of this experiential safety and health training as it can immerse trainees in a near-real simulated work environment and let them experience a serious accident under safe circumstances. This near-real training experience allows trainees to experience the consequence of risky behaviors at work, facilitating the activation of the experiential learning cycle for effective formulation of safety knowledge that is transferable to their work in real life. Additionally, the severe consequence of non-compliance with safety regulations may have a long lasting and impactful effect on their memories, which, in turn, may possibly reduce their risk-taking propensity, leading to a higher motivation for safety compliance and commitment. Detailed comparisons of the training effectiveness between immersive VR training and traditional classroom training should be conducted in future endeavors to provide more concrete evidence for the expected positive effects of immersive VR safety training. Also, other levels of evaluation than just measuring the confidence of the participants after the training should be used, such as investigating the number of working at height accidents and monitoring the participants in their work after the training.

REFERENCES

Angelopoulou, M. V., K. Kavvadia, K. Taoufik, and C. J. Oulis. 2015. Comparative clinical study testing the effectiveness of school based oral health education using experiential learning or traditional lecturing in 10 year-old children. *BMC Oral Health* 15(1), 51.

Burke, M. J., S. A. Sarpy, K. Smith-Crowe, S. Chan-Serafin, R. O. Salvador, and G. Islam. 2006. Relative effectiveness of worker safety and health training methods. *American Journal of Public Health* 96(2), 315–324.

Burke, M., R. Salvador, K. Smith-Crowe, S. Chan-Serafin, A. Smith, and S. Sonesh. 2011. The dread factor. *The Journal of Applied Psychology* 96(1), 46–70.

Christou, C. 2010. Virtual reality in education. In *Affective, Interactive and Cognitive Methods for E-Learning Design: Creating an Optimal Education Experience*, eds. A. Tzanavari and N. Tsapatsoulis, Chapter: 12, 228–243. Hershey, PA: IGI Global.

Donabedian, A. 1966. Evaluating the quality of medical care. *The Milbank Memorial Fund Quarterly* 44(3), 166–206.

Filigenzi, M. T., T. J. Orr, and T. M. Ruff. 2000. Virtual reality for mine safety training. *Applied Occupational and Environmental Hygiene* 15(6), 465–469.

Freina, L., and M. Ott. 2015. A literature review on immersive virtual reality in education: State of the art and perspectives. In *Proceedings of the International Scientific Conference eLearning and Software for Education*, Vol. 1, 133–141. Bucharest: National Defence University (NDU) – Carol I Publishing House.

Gallagher, A. G., N. McClure, J. McGuigan, I. Crothers, and J. Browning. 1999. Virtual reality training in laparoscopic surgery: A preliminary assessment of minimally invasive surgical trainer virtual reality (MIST VR). *Endoscopy* 31(04), 310–313.

Gavish, N., T. Gutiérrez, S. Webel, J. Rodríguez, M. Peveri, U. Bockholt, and F. Tecchia. 2015. Evaluating virtual reality and augmented reality training for industrial maintenance and assembly tasks. *Interactive Learning Environments*, 23(6), 778–798.

Gupta, S. K., D. K. Anand, J. E. Brough, M. Schwartz, and R. A. Kavetsky. 2008. *Training in Virtual Environments: A Safe, Cost-Effective, and Engaging Approach to Training*. College Park, MD: CALCE EPSC Press.

Gurusamy, K. S., R. Aggarwal, L. Palanivelu, and B. R. Davidson. 2009. Virtual reality training for surgical trainees in laparoscopic surgery. *The Cochrane Database of Systematic Reviews*. Hoboken, NJ: John Wiley and Sons, 1–63.

HKU – University of Hong Kong. Department of Industrial and Manufacturing Systems Engineering of the Faculty of Engineering. 2015. *HKU Faculty of Engineering Creates Innovative "imseCAVE" – A High Performance, Low Cost, Virtual Environment for Industrial Applications and Training*. Hong Kong: HKU.

HSE – Health and Safety Executive. 2012. *Health and Safety Training: A Brief Guide*. Bootle, UK: HSE.

Jeelani, I., A. Albert, R. Azevedo, and E. J. Jaselskis. 2016. Development and testing of a personalized hazard-recognition training intervention. *Journal of Construction Engineering and Management* 143(5), 04016120.

Jeelani, I., K. Han, and A. Albert. 2017. Development of immersive personalized training environment for construction workers. *Computing in Civil Engineering*, Proceedings of the ASCE International Workshop on Computing in Civil Engineering 2017, 407–415.

Kim, K., M. Z. Rosenthal, D. J. Zielinski, and R. Brady. 2014. Effects of virtual environment platforms on emotional responses. *Computer Methods and Programs in Biomedicine* 113(3), 882–893.

Kolb, D. 1984. *Experiential Learning: Experience as the Source of Learning and Development*. Englewood Cliffs, NJ: Prentice Hall Inc.

Lau, H., L. Chan, and R. Wong. 2007. A virtual container terminal simulator for the design of terminal operation. *International Journal on Interactive Design and Manufacturing* 1(2), 107–113.

LD – Labour Department. 2019. *Occupational Safety and Health Statistics 2018*. Hong Kong: Labour Department, The Government of the Hong Kong Special Administrative Region.

Lele, A. 2013. Virtual reality and its military utility. *Journal of Ambient Intelligence and Humanized Computing* 4(1), 17–26.

Li, X., W. Yi, H. L. Chi, X. Wang, and A. P. Chan. 2018. A critical review of virtual and augmented reality (VR/AR) applications in construction safety. *Automation in Construction* 86, 150–162.

Lucas, J., W. Thabet, and P. Worlikar. 2008. A VR-based training program for conveyor belt safety. *Journal of Information Technology in Construction* 13(25), 381–407.

Maslen, S. 2014. Learning to prevent disaster: An investigation into methods for building safety knowledge among new engineers to the Australian gas pipeline industry. *Safety Science* 64, 82–89.

Matsas, E., and G. C. Vosniakos. 2017. Design of a virtual reality training system for human-robot collaboration in manufacturing tasks. *International Journal on Interactive Design and Manufacturing* 11(2), 139–153.

Mestre, D. R. 2017. CAVE versus Head-mounted Displays: Ongoing thoughts. *Electronic Imaging* 2017(3), 31–35.

Moshell, M. 1993. Three views of virtual reality: Virtual environments in the US military. *Computer* 26(2), 81–82.

Ong, S. K., and A. Y. C. Nee. 2004. *Virtual and Augmented Reality Applications in Manufacturing*. London, UK: Springer.

Pantelidis, V. S. 2010. Reasons to use virtual reality in education and training courses and a model to determine when to use virtual reality. *Themes in Science and Technology Education* 2(1–2), 59–70.

Perlman, A., R. Sacks, and R. Barak. 2014. Hazard recognition and risk perception in construction. *Safety Science* 64, 22–31.

Porcino, T. M., E. Clua, D. Trevisan, C. N. Vasconcelos, and L. Valente. 2017. Minimizing cyber sickness in head mounted display systems: Design guidelines and applications. In *2017 IEEE 5th International Conference on Serious Games and Applications for Health (SeGAH)*, 1–6, Perth, AU: IFEE.

Pugsley, K. E., and L. H. Clayton. 2003. Traditional lecture or experiential learning: Changing student attitudes. *The Journal of Nursing Education* 42(11), 520–523.

Sacks, R., A. Perlman, and R. Barak. 2013. Construction safety training using immersive virtual reality. *Construction Management and Economics* 31(9), 1005–1017.

Seth, A., J. M. Vance, and J. H. Oliver. 2011. Virtual reality for assembly methods prototyping: A review. *Virtual Reality* 15(1), 5–20.

Swuste, P., and F. van Dijk. 2019. (Post)academic safety and health courses, how to assess quality? In *Occupational and Environmental Safety and Health: Studies in Systems, Decision and Control 202*, eds. P. M. Arezes, J. S. Baptista, M. P. Barroso, P. Carneiro, P. Cordeiro, C. Nélson, R. B. Melo, et al., 785–790. Basel, Switzerland: Springer International Publishing.

Thomsen, A. S. S., D. Bach-Holm, H. Kjærbo, K. Højgaard-Olsen, Y. Subhi, G. M. Saleh, Y. S. Park, et al. 2017. Operating room performance improves after proficiency-based virtual reality cataract surgery training. *Ophthalmology* 124(4), 524–531.

Yuen, K. K., S. H. Choi, and X. B. Yang. 2010. A full-immersive CAVE-based VR simulation system of forklift truck operations for safety training. *Computer-Aided Design and Applications* 7(2), 235–245.

10 People-Oriented Teaching Intervention for Tea Plantation Workers in Assam
A Teaching Intervention Study

Hanan M. F. M. M. Elnagdy

CONTENTS

If you are cold, Tea will warm you,
If you are heated, it will cool you,
If you are depressed, it will cheer you,
If you are excited, it will calm you.

(Glodstone 1809–1898)

10.1 INTRODUCTION

Assam is blessed with tea. On both sides of the Brahmaputra river lies the world's largest tea-growing area. The tea plant type is "Camellia Assamica", a kind of tropical tree plant. Assam is tropical in nature with an annual rainfall of 2,500–3,000 mm between March and September, with daytime temperatures of 35°C–38°C, and humidity levels above 80% around summertime. The combination of humidity and heat creates a greenhouse for rearing the tea plant.

The black tea of Assam is in specific demand on the international market because of its unique flavor and bright liquor. That is due to the nature of the tea tree and loamy fertile soils that help to grow a high-quality tea leaf (Arya 2013). Assam has a harvest schedule of four times per year between March and early December. The final tea product comes from two different methods of production. These two methods are the "orthodox" method and the "crushing, tearing, and curling" method (CTC). The fineness of orthodox leaf tea is considered the geographical trademark of Assam with its golden and chunky appearance (Kadavil 2007).

The total number of tea gardens in Assam is 85,344, including 68,500 small tea growers (Economic survey Assam 2016–2017). The total area under tea cultivation in the state is 312,210 ha, equaling 3,122 km² (Indian Tea Association 2017), with a total production of 676.31 million kg. The total area under tea cultivation for small growers is 117,000 ha, equaling 473.5 km² (IIDCs n.d.), with a total production of 285.24 million kg (Tea Board India 2017–2018).

10.2 DESCRIPTION OF LABOR IN TEA PLANTATIONS IN ASSAM

10.2.1 LABOR FORCE

The labor force of Assam tea plantations represents around 20% of Assam's total population, equivalent to about 6.8 million people. Tea gardens employ about 50% of rural labor. The total daily employment across the state is also estimated to be 50% in tea gardens. Tea labor, therefore, is one of the biggest contributors among the organized workforce to the economy of the state: about 17% of the total workers in Assam are engaged in the tea industry, and around 50% of the total tea plantation workforce are women.

10.2.2 CLASSIFICATION OF TEA PLANTATIONS

Tea plantations are divided into small, medium, and large size according to the Plantation Labor Act of 1951, with small tea plantations ranging from 5 ha to 50 ha, medium-sized ones from 50 ha to 120 ha, and large tea plantations at 121 ha and above. Tea plantations are owned by the government as well as public and private owners (The Assam plantation labor amendment rules (draft) 2001).

The main working areas in the tea industries are the gardens, called the "tea bushes", and the factory. The activities in the garden are plucking, spraying, fertilizing, pruning, and weeding. Factory activities include controlling, observing the production line, lifting, sorting, weighing, and packaging for the final products after processing. The administrative jobs in the tea garden include manager, submanager, welfare officer, teacher, nurse, and doctor.

10.2.3 THE GENERAL HEALTH SITUATION IN TEA PLANTATIONS

The poor standard of living, hygienic problems, bad sanitary systems, and low-quality food are the general problems faced. The workers' livelihood entirely depends on the tea business and on the basic needs provided by planters inside the tea communities. According to the 1951 Labor Plantation Act, the planter should provide the permanent workers with accommodation, water pumps, wells, general bathrooms, and the like. Given the high number of gardens, the facilities of the planters are not enough to deliver a healthy lifestyle for workers. Some tea gardens continue to have insufficient numbers of bathrooms, with some having none, hence "open defecation" occurs especially in medium-sized and small gardens.

Poor drainage systems, allowing clean pumped water to mix with sewer water, cause serious health problems (Kashem 2015). The individual habits of workers contribute to the spread of health problems inside the tea community. Habits of keeping domestic animals inside the houses and unvaccinated pet "dogs" are a conducive environment to the transfer of infectious microbes and worms. Drinking tea with salt (a traditional habit of British colonialists) and alcohol increases workers' susceptibility to high blood pressure (Prasanta et al. 2018). Moreover, alcohol and drugs are common reasons for the violence presently occurring inside the tea communities.

The socio-economic conditions and the unpleasant lifestyle, involving poor hygienic knowledge and practices, cause multiple infections, from moderate to severe ones, including skin infections, diarrhea, intestinal worms, respiratory infections, tuberculosis, and others. Poor nutrition ("malnutrition") leads to stunted growth and cognitive development of children, as well as high rates of morbidity and mortality among the tea workers. Despite the fact that management has provided food at subsidized rates and put some measures in place to improve hygiene, the workers' lifestyles and health situation are not fully improved (Medhi et al. 2006).

10.3 THE WORK ENVIRONMENT IN TEA PLANTATIONS

The work environment in tea plantations includes both the social-economic conditions and occupational safety and health (OSH) aspects inside the tea gardens.

10.3.1 THE SOCIAL-ECONOMIC CONDITIONS

Low wages, low education, unskilled workers, gender inequality, and the poor relationship between workers and management are the common work-related problems in tea plantations.

10.3.1.1   Low Wages

Low wages are the main economic problem for tea workers in Assam. At 145 rupees per working day, approximately US$1.75 according to conversion at that time, their wages are among the lowest in the organized sector. As a result, the tea laborers suffer from poverty, and this has created a vicious cycle carried onto their children as well. The wage payment system does not account for differences in age, gender, permanent or casual workers – they are all entitled to the same daily pay. The factory workers are the only exception; they get more than other workers related to their skills.

10.3.1.2   Education

The education level is very low, with high rates of illiteracy especially among adolescent girls and women. The highest level is grade 10 or 12. This creates another deficiency in tea workers' lives, denying them the opportunity to switch or search for better jobs in other communities, because most of them are unskilled (Baruah and Daimari 2017). The education level has improved in comparison with previous decades, at least in most of the tea gardens medium-sized and above; schools up to grade five have been built inside the tea communities. The new generation of workers' children has more opportunities compared with their parents. Nowadays they can reach university level and hence become professionals in different fields. But still not all of the tea workers' children are covered by the educational efforts introduced by the government and the tea authorities (Sarma 2013).

10.3.1.3   Work Relationships

Weak supervision, poor communication, and unclear job descriptions create misunderstanding, violence, and weak relations between workers and the management. On the other side, the good practice of work relationships between supervisors and tea workers stands as a great motivator for workers to express their opinion towards the maintenance of tea gardens (see Waterman, Chapter 8 in this book). This builds a platform for workers to give feedback on their working conditions. Good relationships are a sign of workers' good performance and the tea gardens' high productivities. Unfortunately, the majority of tea managers neglect their workers' opinions or their participation in management decisions (Goowalla 2012). The government made a step to improve working conditions by appointing a welfare officer based on "The Plantations (Welfare officers) Assam rules 1977". Enhancing worker-management relations is part of the welfare officer's responsibility. He or she is also responsible for undertaking advisory roles on employer matters concerning the provision of particular services such as clear water for drinking, medical facilities, and education facilities, among others. The officer also handles wage problems, sick leave issues, annual leave, and sickness benefit.

10.3.1.4   Gender Inequality

Above 50% of workers on tea plantations and around 70% of casual workers are women. In some tea gardens, this has given women a chance for their voice to be considered in decision-making (Gurung and Mukherjee 2018). Women working in tea gardens "burn out" because of the work intensity and the housework responsibilities that come along with less support from their partners. They are more affected by the inadequate work conditions than men, have a higher percentage of illiteracy, suffer from poor health facilities especially during pregnancy, and at times are subjected to home violence. There are also cases of rape and early marriages (Antich 2016).

10.3.1.5 Health Care and Work-Life Balance

The planter supports the tea garden workers with only minor hospital or clinic diagnosis and medication; however, for specialized medical attention, tea garden families are referred to government hospitals. In terms of work-life balance, workers have to spend ten hours per day at work, meaning they have about six hours left to catch up with their family, engage in some social activities, and take care of their children. This creates a serious challenge for them to strike a balance between work and their personal lives.

10.3.1.6 Job Security

Tea garden workers suffer cutoffs that result from low productivity and high costs of production leading to declining prices of tea. Some of the workers' jobs are seasonal and insecure, as planters are unable to pay a big labor force with no justification, especially during off season. Casual workers only get paid during the high season of tea cultivation, and they must arrange their own accommodation, except for their health needs, which are covered.

10.3.2 Occupational Safety and Health in Tea Plantations

It should suffice to mention that the workers in tea plantations are exposed to all kinds of hazards, that is to say, chemical hazards, biological hazards, and physical hazards, including agronomic hazards, solar radiation and fire exposure, and psychosocial hazards (Borgohain 2013). Cultivation starts in March and continues until early December, a period of summertime characterized by high temperatures and high humidity. In the plucking period, women in particular are exposed to heat and direct sun radiation from 6 a.m. to 4 p.m. all through the workday. The heat and high humid conditions cause sweating and heat inflammation that increases microbe and germ infections among workers. The forceful movement and carrying of heavy awkward loads increase musculoskeletal injuries among tea workers (Fathallah 2010). The baskets and nets used to collect tea leaves are designed to hang on the head and rest on the back, leaving workers' hands free to pick the tea leaves. The basket or net weight can reach up to 40 kg. The baskets' design and weight, as well as the repeated movement while plucking tea, lead to significant injuries of the neck muscles, skull, shoulders, and hands. The bending and lifting while carrying the heavy baskets or nets are additional reasons for musculoskeletal injuries especially back problems (Figure 10.1).

Women can be poisoned or contract serious health problems from their indirect exposure to pesticides on the tea leaves. The pesticides can be absorbed by the skin or by eating with unwashed hands, or by rubbing the eyes with contaminated hands. Women in tea bushes wear slippers; their feet easily get exposed to bites from different insects and snakes. There are also risks of hands and feet being cut by sharp

FIGURE 10.1   Baskets and regular posture while plucking tea.

edges of tree branches and leaves, injuries during uprooting, clearing, and pruning, in addition to the slippery falls, sprains, and fractures (ILO 2004).

Men are responsible for spraying the pesticides and fertilizers and hence have more direct exposure to chemical hazards. Most of the time, no personal protective equipment is used, meaning their hands, arms, legs, and feet are often unprotected. In most if not all situations, workers do not use protection masks while spraying, letting the pesticides affect their lungs. Workers are affected by chemicals causing some cases of toxicity, health problems in the respiratory system, skin and eye inflammations, and rare problems of the cardiovascular system. When it comes to plucking, men are exposed to the same risks as women; 90% of male workers work in the tea factory, exposed to risks caused by old production lines or unguarded belts. The bending and lifting of products from one stage to another, especially after sorting and packaging, causes back pain and injuries to shoulders, arms, and legs (Gupta and Dey 2013). Workers perform their jobs in a poorly ventilated condition, particularly in the sorting area, due to the presence of suspended dust and very fine particles in the air from the final tea product. This can be a source of asthma or respiratory problems. Male workers are exposed to high temperatures exceeding 80°C in the dry stage during long working hours. This can cause burning, fatigue, sweating, and dehydration.

The housekeeping workers face similar hazards as tea factory and tea field workers. Biological hazards can arise from the working conditions and the lifestyle of workers, causing serious health problems such as diarrhea, skin infection, poison, fever, worm intensities, among others (Govindan Kutty and Priyadarshan 2018).

Work-related stress in tea plantations arises from several factors: pressure to collect certain amounts of tea leaves to get the normal salary per day, job insecurity – especially for casual workers, overload during the high season, and poor working conditions, as explained earlier (Mahalakshmi and Mageshkumar 2017). Work-related psychosocial problems also emerge in some other areas, including mobbing and bullying among tea workers and conflicts with the administration.

Workers' lack of awareness of occupational safety and health and a deficiency of OSH skills increases the risk of their involvement in accidents or incidents. In fact, workers with insufficient OSH awareness and skills pose potential hazards to the entire community (Laberge et al. 2014).

10.4 INTERVENTION APPROACH

Intervention can be simply defined as a combination of actions aimed at emphasizing the specific problems involved and producing specific changes in behavior at both the individual and the community level (British Psychological Society 2011). There are several types of interventions, including therapeutic intervention (Beresford et al. 2018), psychosocial intervention (England et al. 2015), and teaching intervention (Machera 2017), among others.

Teaching intervention is a common type of intervention that focuses on enhancing the level of knowledge of specific groups of people in order to motivate their behavioral change. It is also an active technique that can be applied to individuals with different backgrounds, knowledge, gender, and level of education. A teaching intervention covers a range of different methods that allows for selecting a method that best suits a specific group related to a specific aspect. The orientation process takes place during the implementation of the intervention, which increases interaction and knowledge sharing, thus developing better relationships.

The design of the teaching intervention covers the following questions (Goldenhar et al. 2001):

- What are the characteristics of the target group?
- What are the existing occupational safety and health problems in their work environment?
- What changes are necessary to improve the occupational safety and health culture in the target group?
- What are the main obstacles to applying the teaching intervention?
- What are the best tools for the transfer of safety and health knowledge in the target group?
- What is a suitable way to increase the potential of change in the lifestyle of the people and the prevention culture at the tea plantation?
- What is a suitable technique to evaluate the enhanced level of OSH knowledge in the target group and their ability to adopt the change?

10.4.1 THE PURPOSE OF THE TEACHING INTERVENTION

The tea industry in Assam is a booming agro-business employing the highest proportion of workers in the state (54.8%). The workers in tea plantations urgently need (interactive) measures to enhance their safety, health, and wellbeing at work and in their life. Training programs on OSH issues for workers in tea gardens are rare or nonexistent. For that reason, the training program proposal was necessary and timely to help sensitize workers on OSH practices. The program was made flexible to match workers' capabilities. It was also designed to be as simple as possible, giving direct and short messages for workers to understand and interact easily (O'Connor et al. 2014; Weber 2016).

10.4.2 IN ADVANCE OF THE INTERVENTION

To set clear goals and objectives for the intervention, the true prevailing situation in the field was analyzed during a field visit. The visit covered all the working areas in the tea plantation, including tea bushes and factories. It was also extended to the tea workers' communities commonly known as "workers' lines". The information collected during the field visit helped to give a clear understanding of the work hierarchy in the organization of the tea plantation. The information was collected through visualization, general conversations, and communication with the workers, supervisors, and administrative staff. The information covered aspects such as working conditions, working hours, and workers' lifestyle. It is important to know that the labor management is under the *sardars* ("the head of workers who report to supervisor"). There are *sardars* for female workers and *sardars* for male workers. The *sardars* themselves are supervised by the managers, assistant managers, and welfare officers of the tea garden. This keeps the mutual loyalty intact in the tea industry. This loyalty is a strong force of motivation and satisfaction among the tea laborers.

The visit was carried out during the 2016 tea season and covered seven tea plantations, five of which had factories. The visited tea plantations were located in three districts of Assam: Dibrugarh, Sivasagar, and Tinsukia. The conversation with workers was challenging because they could not speak English. The interpretation was done with the help of the assistant manager, the welfare officer, or a student from Dibrugarh University. The visit covered the daily work cycle of workers in the tea plantation. The working hours in factories are different: During the high season from March to September, they work 24 hours in three shifts. In some tea factories, they work in two shifts only. The factories' time schedule changes at the end of the season with one shift after three days due to the reduction of the collected tea.

The visit revealed the level of occupational safety and health in these tea gardens. The OSH level ranged from very low to moderate. The differences in OSH level were related to the size of the tea gardens and the presence of a welfare officer. The level also depended on the extent to which the owner complied with his obligation to OSH legislation. Some gardens adapted standards related to child labor, environmental impact, fire/explosion protection, or chemical exposure protection. No OSH training had been carried out in most of these tea plantations so far. Even communication about the hazards related to their workplaces was rare in the past.

10.4.3 THE DESIGN OF THE TEACHING INTERVENTION

Dibrugarh district was chosen as the district for the teaching intervention to be implemented. It is considered as the district in Assam with the highest number of tea gardens and tea production. The design of the teaching intervention focused on medium-sized tea gardens with no more than 500 workers or a single division with approximately 400 workers in case of a large tea plantation.

The aim of the intervention was to make the behavior of the workers safer and healthier and to improve the relationship between the workers and their managers, specifically in terms of OSH behavior. The intervention had two objectives: the first was "to come up with a simple and direct orientation for workers to increase their knowledge about OSH issues"; the second was "to enhance personal behaviors with regard to avoiding risks in the workplace". To achieve these objectives, the intervention consisted of two measures. The first measure was a campaign to encourage good personal hygiene and to use personal protective equipment in a proper way (ILO 1991; WHO 2009). The second measure was teaching workers about good working postures in different positions, such as heavy loading, bending, lifting, and basket/net carrying (Kawakami et al. 2006; Niu and Kogi 2012).

There were several obstacles the teaching intervention had to cope up with. The first obstacle was the language barrier, because none of the workers knew English, and some did not know "Assamese", the main language of Assam state. The second obstacle was a lack of acceptance by the managers. The language barrier was a big problem not only in terms of encoding and transferring messages but also because workers lost interest in participating due to the fact that they could not understand or share during the activity. In pursuit of trying to find a simple way to communicate with the workers, new questions came up, such as: How do we communicate what? Which illustrative tools shall be used in the intervention? What are suitable ways to encourage the workers to participate? For these matters, the design of the intervention was built up on the basis of "less talk" and more non-verbal activities and body movements. The part involving verbal interactions was designed for critical situations only and carried out in sessions with interpretation, with the help of a volunteering student of Dibrugarh University.

The chosen method for the teaching intervention was based on colorful clear images and posters that easily illustrated safe and unsafe positions, simulating bad postures and comparing them with the good ones. The teaching intervention included games like play ball, an exercise with raising red and green cards, and others. The idea was to motivate the workers by involving them in the activities, like sharing in games, simulating the postures, pointing on photos, and others.

The schedule of the intervention was determined by the information about the work cycle received during the visit. The most suitable time for the implementation was after working hours, particularly on Saturday or Sunday. However, the chosen day depended on workers' preferences. The time plan of the intervention was three contact sessions with the workers. Each session lasted one hour and a half. The first and second contact were one week apart; the third followed two or three months later. The first two sessions covered the two measures of the intervention. The third session was to evaluate whether there was any change in workers' behavior after the intervention. As part of the design specification, the number of participating workers was not to be less than 30% of the total workers in the tea gardens.

Regarding the second obstacle – the lack of acceptance by the management – the aim of the intervention was to encourage the managers to accept the implementation of the teaching intervention inside their tea garden. To achieve this, the intervention was designed in a way to avoid any conflict with the management. The managers were skeptical and unconfident about giving the team permission to enter their tea gardens and to conduct the intervention activities in there. An alternative was found by going to the tea communities' representatives and accommodation lines of tea workers while conducting the intervention activities from there. The tea communities were successful entry points for conducting the teaching interventions.

10.4.4   THE IMPLEMENTATION OF THE TEACHING INTERVENTION

The teaching intervention was conducted in three tea gardens, with 40 to 70 tea workers participating in each tea community. Inside the communities, there were some elders and school children. More women participated in the intervention than men. The total number of workers from the three tea estates and their participation and gender details are shown in Table 10.1. The table also shows the number of children and elders that attended in various tea estates.

The first period of implementation was to introduce the purpose of the teaching intervention and to show how the implementation activities will go on. The communication during this introductory session was supported by interpretation; after that, a mixture of verbal and non-verbal communication was used. The second session covered the first measure of the intervention, including important tips for all workers, elders, and children, for instance, with regard to hand washing and feet protection.

During the second visit a week later, the intervention focused on wrong postures while lifting and bending, as well as on taking caution of fall hazards when working

TABLE 10.1
Worker Numbers, Participants, and Gender

The tea estate	Gender divisions	Permanent workers	Casual workers	Participant workers	Other participants	
					Elders	Children
Bihaiting	Women	199	170	44	7	≈ 20
	Men	168	100	19		
	Total	367	270	63		
Lepatkata (one division)	Women	140	167	32	4	≈10
	Men	97	20	21		
	Total	237	187	53		
Durgapur	Women	123	103	28	3	≈15
	Men	27	22	19		
	Total	150	125	47		

in the tea garden or in the factory. Workers were taught the right postures when holding the baskets and nets while plucking. During the second visit, emphasis was put on personal protective equipment that should be used during the spraying of pesticides and fertilizers.

The third visit was made after three months. The main purpose was evaluating the two measures taught during the previous visits. It focused on checking whether workers memorized the lessons correctly and integrated them into their day-to-day work routines. A measure of the change in the behavior of the participating workers was evaluated. The activities were performed using the technique explained in the design part (Figures 10.2 and 10.3).

10.4.4.1 Intervention Measures Part I: Safety and Health Tips

The teaching intervention measures covered the critical hazards and health problems that workers are exposed to every day in their work or inside their communities. These hazards and problems can easily be addressed and solved to avoid or eliminate the risk. The first tip was mainly about washing hands to enhance hygiene as a safety and health measure. Workers were taught how to wash their hands with clean water and soap.

Using personal protective equipment was the second message during the implementation of the teaching intervention. There was a need to find suitable solutions to protect workers from chemical hazards, infections, and toxicity. These solutions had to be inexpensive and easy to adopt. While spraying, the men should wear clothes covering their whole body, including their faces to avoid inhaling the sprayed fertilizer and pesticides. The covering cloth can be a normal piece, wet with clean water. This can be of much better help and at the same time less costly. However, if management supplied safety overalls and proper face masks, this would be a better option. Workers' hands should be covered by gloves and their feet by closed shoes, not slippers. This applies not only to the men but also to the women and children. The aim is to decrease as much as possible the risk of contracting skin infections from worms and insects, especially in tea communities with poor hygiene. The solutions

FIGURE 10.2 Card raising during the teaching intervention.

FIGURE 10.3 Posture simulation during the teaching intervention.

mentioned above are not only effective but can also be applied using materials available at the nearest market at a reasonable price.

10.4.4.2 Intervention Measures Part II: Good and Bad Postures

The last measure of the intervention was about good and bad postures, covering a huge area of the activities on the tea plantation. Wrong postures when bending and lifting were demonstrated clearly during the visits in the tea garden and inside the factory.

Unfortunately, attempts to convince the administration of reducing the weight of the tea basket or modifying the production line so that workers become less exposed to the risk were unsuccessful, but alternative solutions to decrease the hazard were presented. During the intervention, postures involving both bad and good positions were simulated. Photos of real positions in different tea plantations were used, and workers expressed themselves by pointing at photos that depicted their daily postures. During the intervention, they learned the correct postures of lifting, bending, bending while lifting, and other positions. With most fields employing over 80% of their tea workers as pluckers, the risk from the hanging tea basket or net is considered one with the highest exposure. This fact called for more attention while conducting the intervention. To reduce the risk, workers were introduced to two modifications: modifying the hanging position of the tea basket or net and reducing the amount of heavy-weight lifting.

Modifying the hanging position for the tea basket or net was made so easy it did not even require replacing the used basket. Workers only had to change the hanging part and make it suitable for shoulder hanging, much like a backpack. This position is safer and healthier, because it no longer affects the neck and the skull and reduces the bending level. However, carrying a heavy weight on the shoulders and the back is another hazard. To remedy this second hazard, workers were later encouraged to empty their basket or net on a regular basis, at least two times in both the morning and afternoon shift. The recommended weight is between 5 kg and 7 kg on the shoulders, as compared with 25 kg as it is for the current situation. Reducing the weight and emptying the baskets and nets regularly was an effective solution in reducing the current potential hazard.

10.5 EVALUATION OF THE TEACHING INTERVENTION

The third visit to every division or medium-sized tea garden was an evaluation visit. This visit was conducted after three months or more following the implementation of the teaching intervention. The goal of the evaluation was to find out whether the intervention enhanced the intended level of OSH knowledge among the tea workers, whether there was any change in their behavior, and whether the relationship between the workers and their managers had improved. A check-up was made to find out whether they still remembered the core messages from the previous intervention. Workers were asked what exactly they remembered from the implementation, and whether there was any improved behavior among the tea workers. The evaluation was conducted by asking direct questions through the interpreter and by playing some games that called for the workers' interaction. The questions were easy to understand and direct so that workers could respond easily. From the three plantations, 75 tea workers responded to the questions. Workers kept excellent memory of the simulation postures, games, and posters (100%). Around 92% of the workers remembered the core messages of the two teaching interventions. It was found that a reasonable percentage of workers (about 71%) understood the messages and explained why they were important for their lives.

A weak point was behavioral change. Among all participants, only nine (seven women and two men) adopted new OSH practices in their lives. These 12% of participants reported a change in their behavior both in their personal and working life.

10.6 CONCLUSIONS

The teaching intervention produced some remarkable lessons. The first lesson is the power of "occupational safety and health language". OSH is a unique language that is wordless, simple, direct, easy to understand, attractive, and effective. There are a lot of international examples showing the elements of OSH language, which include sign marks, movies, and animations. "OSH language" was a means to overcome language barriers with the tea workers while implementing the teaching intervention. Photos, well-designed posters, and posture simulations helped trainers communicate the OSH messages with less effort because they were non-verbal. The non-verbal messages were easily understood by the tea workers, easy to memorize, and finally improved workers' OSH knowledge.

The second lesson was: "Bring the service to the door of the target group". The marginalized workers never had the chance to participate in any OSH training program before this teaching intervention. Tea workers do not easily leave their duties during workdays, even if it is for training. Managers and supervisors are focused on productivity, particularly in the high season of tea cultivation. For this reason, they do not give more interest or care about OSH training, which leaves a deficiency in the workers' OSH knowledge and exposure. Through the teaching intervention, we had the chance to introduce the OSH training in tea communities and in the tea workers' houses. The workers were surprised and excited, because they felt capable of expressing themselves while staying in their homes. They were happy to share their experience during the intervention activities and were able to accept the OSH measures. They appreciated the efforts taken in visiting their communities and conducting the intervention in their setting. They were also more willing and waiting for the second meeting to learn new OSH messages.

The third lesson was that "a teaching intervention is a remarkable approach in itself" – a technique that is easy to customize in order to serve different categories of people. The variety of methods in teaching interventions makes the approach more flexible to adapt to the different capabilities and different backgrounds of individuals or groups. The implementation sessions were a success, particularly in terms of message delivery and enhancement of OSH knowledge. Workers received critical information about their safety and health at the workplace through simple ways such as playing games, body language, and simulation.

Finally, the intervention promoted workers' knowledge and enhanced their OSH and livelihood competences. Tea workers are unique in their work-life and lifestyle. They do have unique skills, even though they are considered unskilled workers in the labor market.

The teaching intervention objectives focused on workers' development. The goal was to improve the OSH level inside the tea industries with respect to both work and community life through the change of workers' behavior by triggering their thoughts towards OSH. This exercise involved introducing prevention culture into their lives, giving them the confidence to express themselves, and demonstrating suitable and simple solutions to be adopted easily with less effort and in line with their budget.

ACKNOWLEDGMENT

The author is grateful to Bailung Himangshu for his support in the selection of tea plantations used for carrying out the teaching intervention, and for mobilizing and coordinating the support of the students of the Geographical Institute, who were a great help in the translation.

REFERENCES

Antich, V. W. 2016. *Analysis of the Impact of WASH Interventions on Women and Adolescent Girls' Well-Being in the Tea Plantations of India: A Comparative Case Study of Ghoronia and Singlijan Tea Estates in Dibrugarh, Assam.* Valencia: University of Valencia.

Arya, N. 2013. Indian tea scenario. *International Journal of Scientific and Research Publications (IJSRP)* 3(7), 46–53.

Baruah, P., and M. Daimari. 2017. Education of tea tribe children: A case study of Udalguri District of Assam. *International Journal of Advances in Social Science and Humanities* 5(5), 26–35.

Beresford, B., S. Clark, and J. Maddison. 2018. Therapy interventions for children with neurodisabilities: A qualitative scoping study. *Health Technology Assessment* 22(3), 1–150.

Borgohain, P. 2013. Occupational Health hazards of tea garden workers of Hajua and Marangi tea estates of Assam, India. *The Clarion* 2(1), 129–140.

British Psychological Society and the Royal College of Psychiatrists. 2011. *Psychosis with Coexisting Substance Misuse – Assessment and Management in Adults and Young People.* NICE Clinical Guidelines 120. National Collaborating Centre for Mental Health (UK). Leicester, UK: British Psychological Society.

Economic survey Assam. 2017. *Economic Survey of Assam 2016 – 17.* Directorate of Economics and Statistics. Dispur, India: Government of Assam.

England, M. J., A. Stith Butler, and M. L. Gonzalez. 2015. *Psychosocial Interventions for Mental and Substance Use Disorders: A Framework for Establishing Evidence-Based Standards.* Washington, DC: National Academies Press.

Fathallah, F. A. 2010. Musculoskeletal disorders in labor-intensive agriculture. *Applied Ergonomics* 41(6), 738–743.

Goldenhar, L. M., A. D. LaMontagne, T. Katz, C. Heaney, and P. Landsbergis. 2001. The intervention research process in occupational safety and health: An overview from the National Occupational Research Agenda Intervention Effectiveness Research team. *Journal of Occupational and Environmental Medicine* 43(7), 616–622.

Goowalla, H. 2012. Labour relations practices in tea industry of Assam – with special reference to Jorhat District of Assam. International Organization of Scientific Research (IOSR). *Journal of Humanities and Social Sciences (IOSR-JHSS)* 1(2), 35–41.

Govindan Kutty, K., and S. Priyadarshan. 2018. Study on occupational health hazards and the safety measures taken by the tea plantation workers of Kerala. *International Journal for Science and Advance Research in Technology (IJSART)* 4(4), 21–29.

Gupta, R., and D. Sanjoy. 2013. Assessment of safety and health in the tea industry of Barak Valley, Assam: A fuzzy logic approach. *International Journal of Occupational Safety and Ergonomics: JOSE* 19(4), 613–621.

Gurung, M., and S. R. Mukherjee. 2018. Gender, women and work in the tea plantation: A case study of Darjeeling Hills. *The Indian Journal of Labour Economics* 61(3), 537–553.

IIDCs – Industries & Commerce. Government of Assam, India. n.d. *Tea Industries.* https://industries.assam.gov.in/portlet-innerpage/about-tea-industries (accessed July 1, 2019).

ILO – International Labour Organization. 1991. *Safety and Health in the Use of Agrochemicals: A Guide.* Geneva: International Labour Office.

ILO – International Labour Organization. 2004. *Hazardous Child Labour in Agriculture "Tea".* Safety and Health Fact Sheet. International Programme on the Elimination of Child Labour. Geneva: International Labour Office.

Indian Tea Association. n.d. *About Tea.* https://www.indiatea.org/tea_growing_regions (accessed July 1, 2019).

Kadavil, S. M. 2007. *Indian Tea Research Study.* https://www.somo.nl/wp-content/uploads/2007/01/Indian-Tea-Research.pdf (accessed July 1, 2019).

Kashem, A. 2015. *Health and Sanitation Behaviour of the Tea Garden Labourers: Crises and Deprivation.* Shahjalal University of Science & Technology. doi:10.2139/ssrn.2560822.

Kawakami, T., S. Arphorn, and Y. Ujita. 2006. *Work Improvement for Safe Home (WISH): Action Manual for Improving Safety, Health and Working Conditions of Home Workers.* Geneva: International Labour Office.

Laberge, M., E. MacEachen, and B. A. Calvet. 2014. Why are occupational health and safety training approaches not effective? Understanding young worker learning processes using an ergonomic lens. *Safety Science* 6, 250–257.

Machera, R. P. 2017. Teaching intervention strategies that enhance learning in higher education. *Universal Journal of Educational Research* 5(5), 733–743.

Mahalakshmi, A., and D. Mageshkumar. 2017. Occupational Health hazards of tea plantation workers of Anamallais (Valparai), Coimbatore district, TamilNadu. *International Journal of Research in Engineering and Social Sciences (IJRESS)* 7(1), 142–149.

Medhi, G. K., N. C. Hazarika, B. Shah, and J. Mahanta. 2006. Study of health problems and nutritional status of tea garden population of Assam. *Indian Journal of Medical Sciences* 60(12), 496–505.

Niu, S., and K. Kogi, eds. 2012. *Ergonomic Checkpoints in Agriculture. Practical and Easy-to-Implement Solutions for Improving Safety, Health and Working Conditions in Agriculture.* International Labour Organization (ILO) in collaboration with the International Ergonomics Association (IEA). Geneva: International Labour Office.

O'Connor, T., M. Flynn, D. Weinstock, and J. Zanoni. 2014. Occupational safety and health education and training for underserved populations. *HHS Public Access* 24(1), 83–106.

Sarma, G. 2013. A case study on socio-economic condition of tea garden laboureres – Lohpohia Tea Estate of Jorhat District, Assam. *Pratidwhani the Echo* 1(3), 50–60.

Tea Board India. 2017–2018. *6th Annual Report of the Tea Board 2017–2018.* http://www.teaboard.gov.in/pdf/64th_Annual_Report_2017_18_pdf4214.pdf (accessed July 1, 2019).

The Assam plantation labor amendment rules (draft). 2001. https://labourcommissioner.assam.gov.in/sites/default/files/Plantations%20Labour%20Rules.pdf (accessed July 1, 2019).

Weber, C. 2016. *Foundational Principles of Instruction and Intervention Systems.* Rexford, NY: International Center for Leadership in Education. Boston, MA: Houghton Mifflin Harcourt.

WHO – World Health Organization. 2009. *Guidelines on Hand Hygiene in Health Care.* First Global Patient Safety Challenge. Clean Care Is Safer Care. Geneva: WHO Press.

11 New Competences of Safety Professionals
A Comprehensive Approach

Rüdiger Reitz and Martin Schröder

CONTENTS

11.1 SAFETY PROFESSIONALS IN GERMANY: A BRIEF INTRODUCTION

In Germany, the Occupational Safety Act (*Arbeitssicherheitsgesetz*) requires every organization to employ or contract safety and health experts. The law calls for two kinds of experts: medical specialists (*Betriebsärzte*) on the one hand and safety professionals (*Fachkräfte für Arbeitssicherheit*, commonly referred to as "Sifa" in German) on the other. The deployment time that must be planned for these experts to safeguard occupational safety and health depends on multiple factors, such as the number of employees and the degree of risk at a given workplace. Their duties include:

- advising management with respect to implementing a comprehensive plan for the assessment of working conditions;
- offering support and advice when carrying out the assessment of working conditions;
- providing support with respect to basic measures to design working conditions;
- providing advice for choosing and implementing suitable protective measures; and
- providing support with integrating occupational safety into corporate organization and management structures.

To be able to perform these duties in line with legal requirements, safety professionals must have defined professional qualifications such as a master craftsperson qualification or a university degree in engineering, as well as a minimum of two years of work experience. Moreover, the law requires safety professionals to complete an appropriate training program. Such programs have been available in Germany since 1979 as part of a systematic training scheme. Following a revision in the early 2000s, the most recent revision has led to the third redesigned version of the program.

Aside from the statutory accident insurance institutions, which are legally required to offer such courses, providers of these training programs include universities and commercial providers specifically accredited for this purpose. Overall, there are about 40 providers, training more than 3,000 safety professionals every year. Most of the programs are designed as part-time courses for working professionals and take about one and a half years to complete. After successfully graduating from the program, participants receive a certificate of expertise.

11.2   DESIGN AND DEVELOPMENT OF THE SIFA TRAINING PROGRAM

11.2.1   TRADITIONAL TRAINING APPROACHES

Since 1979, two different training models for prospective safety professionals have existed in Germany. These two models were to be standardized and thoroughly modernized to reflect the most recent findings of adult education research.

The training program to become a safety professional in Germany consists of three levels (BMA 1997; DGUV 2012).

- Level I: Basic training;
- Level II: Advanced training; and
- Level III: Customized training for specific industries.

The training curriculum for levels I and II is the same throughout Germany, whereas level III contents are determined by the relevant statutory accidence insurance institution in response to the concrete needs of a given industry.

11.2.2   THE CALL FOR REDESIGNING THE SIFA TRAINING PROGRAM

In 2009, the bodies of the German Social Accident Insurance (DGUV) officially ordered a redesign of the Sifa training program. In 2011, a basic training model with modern educational guidelines was presented (DGUV 2011a). The training model follows a systemic-constructivist approach known in Germany as "facilitated learning", or *Ermöglichungsdidaktik*, a term coined by education researcher Rolf Arnold.

> Safety professionals are trained according to an action-oriented approach that focuses on the development of demand-specific competences. The prospective safety professional should learn to act and collaborate with other people in a self-directed and responsible manner in order to further develop occupational safety and health in their organization.
>
> **(Arnold 2007; DGUV 2011a, author's translation)**

Between 2012 and September 2019, the training model was translated into concrete formats and media. The goal was to create a blended learning solution enabling participants to develop the required competences in an individualized, self-directed, and sustainable manner. The curriculum was adapted to the current state of technology and, most importantly, to the needs of today's workplace. The role of the safety professional, too, has changed since the early 2000s, regardless of legal requirements: Formerly primarily a source of technical expertise, today's Sifa has evolved into an advisor who takes a holistic view of company organization, the prevailing culture of prevention, and policies to protect the health of the workforce.

But the focus is not only on the hazards that employees encounter at the workplace. From the perspective of demands and resources, the modern safety professional looks at "how external influences protecting or promoting the health of the workforce can serve as resources that have a positive impact on health and help to strengthen the individual's internal health resources" (Sifa knowledge module 4.05, "Systemisches Anforderungs-Ressourcen-Modell", 5).

11.2.3 EXCURSION: FINDINGS FROM THE LONG-TERM SIFA STUDY

A longitudinal study of the work of safety professionals (Hamacher et al. 2013) confirmed the changes in the roles they serve today, providing valid figures on the effectiveness of safety professionals at companies and public authorities. Between 2004 and 2011, some 9,000 data sets were collected via multiple surveys at roughly two-year intervals. In addition to some 4,000 safety professionals, each survey covered about 500 managers/senior officers, company doctors, and works council members to evaluate how the effectiveness of safety professionals is perceived not only by themselves but also by other relevant stakeholders.

One discouraging result of the study was that roughly one-third of all safety professionals found that their work had little or no effect at all so far.

The study revealed that there is no single factor that makes the work of safety professionals effective; rather, there are multiple factors working together to increase a safety professional's effectiveness. The core outcome of the study is a model of highly interconnected causal relations that identifies the various factors at work and how they influence each other. In particular, these factors include the conditions under which safety professionals deliver their services for a given organization, for instance, as in-house specialists or as external service providers. The structural setup in a given industry also plays a role. Likewise, a company's size, type, and economic performance have an impact on how occupational safety and health requirements are met. And last but not least, corporate culture and corporate goals serve as a crucial foundation for the effectiveness of safety professionals. Without the support or commitment of management, the extent to which a Sifa can make a difference is very limited.

At the personal level, the study found three factors to influence the effectiveness of a safety professional:

1. personal prerequisites;
2. work methods; and
3. attitude, perceptions, and motivations.

Among these three factors, "work methods" has the strongest impact on the work of safety professionals. Those who exert systematic and conceptual influence on an organization's safety- and health-related management processes are demonstrably more effective than those who only deal with the technical aspects of occupational safety.

Furthermore, the study investigated individual aspects with respect to the effectiveness of comprehensive, modern prevention activities. Here are some selected results:

• One crucial factor in terms of effectiveness is whether a corporate master plan for the assessment of working conditions was developed and implemented.

- A comprehensive assessment of working conditions must include psychological stress and unfavorable mental strain.
- In terms of prevention, one area with substantial potential for improvement is the humane organization of work.
- The more intensely a safety professional addresses the organization of occupational safety, the more effective their efforts will be.
- Collaboration with many partners and the quality of that cooperation has the strongest effect on the organization of occupational safety.
- One essential factor impacting on the safety professional's work and effectiveness is the prevailing culture of prevention within a company.

Regarding the competences required of a safety professional, the study found that a systematic and methodical approach, the ability to collaborate with others, and self- and other-oriented social competences are key prerequisites for acting effectively within the organization.

The long-term Sifa study, which was completed in 2012, thus supports the fundamental decision to move towards a competence orientation in the redesigned Sifa training program.

11.2.4 CONCLUSION: THE SIGNIFICANCE OF COMPETENCE ORIENTATION

Aside from the general intra- and supra-company conditions and circumstances that shape the safety professional's work, the development of personal (action) competences is crucial for effective prevention work. Against this background, the redesigned training model puts particular emphasis on acquiring these competences, not least to strengthen and positively influence the outcome of the training: the effectiveness of the safety professional within their organization.

11.3 FROM ACTIVITIES TO A COMPETENCE PROFILE

11.3.1 THE REGULATORY FRAMEWORK GOVERNING THE DUTIES OF SAFETY PROFESSIONALS

In Germany, the deployment of safety professionals is regulated by the Occupational Safety Act (*Arbeitssicherheitsgesetz*):

> Safety professionals shall support the employer in all matters of occupational safety and health in connection with accident prevention and all issues of safety at work, including measures to promote the humane working conditions.
>
> **(Arbeitssicherheitsgesetz § 6, sentence 1, author's translation)**

This general regulation is specified by a legal provision of the statutory accident insurance institutions, DGUV Provision 2 (DGUV 2009; DGUV 2011). In the concrete implementation of the training model, DGUV Provision 2 served as a key foundation for the redesign of the Sifa training program. On the basis of this provision, the competence profile of safety professionals was further enhanced and specified. The Sifa's basic responsibilities towards the insured companies are always the same, regardless of differences in terms of industry and the scope of work. With regard to basic supervision provided to companies, Annex 3 of DGUV Provision 2 lists nine areas of activity for safety professionals (and occupational physicians):

1. support with risk assessments (assessment of working conditions);
2. support with basic prevention activities focusing on the work environment;
3. support with basic prevention activities focusing on individual behavior;
4. support with the creation of a suitable structure and integration into managerial activities;
5. investigations following events;
6. provision of general advice to employers and managers, workforce representatives, and employees;
7. documentation, obligatory reporting;
8. involvement in enterprise meetings; and
9. self-organization.

All nine task areas are described in more detail in the provision. In addition to the nine areas of activity in basic supervision, the provision identifies four areas that may call for company-specific supervision. These include:

1. regularly occurring company-specific accident and health hazards; human factor requirements for work design;
2. changes to working conditions and organization;
3. external developments that have a specific influence on the situation in the enterprise; and
4. campaigns, programs, and activities in the enterprise.

In the training model, these task areas served as the basis for deriving the job profile of safety professionals, which lists some 200 individual tasks (DGUV 2011a, Annex 3).

According to German law, safety professionals serve in a staff function, not in a line function. Their job is to provide support and advice to employers and managers. As a consequence, the key question when developing a competence profile for safety professionals is identifying the skills and abilities they must have to be able to effectively advise, support, and motivate within their organization.

11.3.2 Factors That Shape Actions: Competence Areas in Sifa Training

Every participant in the Sifa training program already has a wide range of different abilities and skills (competences) from their prior vocational training and career under their belts. The Sifa training program and the Sifa competence profile build on these existing abilities and skills, especially on participants' knowledge and understanding of company procedures and their "personal attitude". Yet participants usually do not have sufficient knowledge of and experience with specific occupational safety and health issues. It is against this background that the course develops the specific competences of a safety professional, based on a combination of the four basic competences: technical, methodological, social, and personal competence (Weinert 2001).

Building on that distinction, the Sifa training model describes the competences as follows:

- Know-how;
- Dealing with others;
- Dealing with oneself; and
- Stance/attitude.

These competence areas are each conceived in multiple dimensions as "factors that shape actions", with close interrelations between the various subcompetences (DGUV 2011a).

The Sifa job profile was then used to derive the target competences for a safety professional. To do this in a structured manner, the KODE competence atlas by Volker Heyse and John Erpenbeck (2009) was chosen as a grid and supplemented with two additional competences that matter for safety professionals: role awareness and self-reflection. The core of the Sifa training program thus consists of 13 basic competences.

- Know-how:
  - Technical expertise
  - Systematic and methodical approach
- Dealing with oneself:
  - Initiative
  - Willingness to learn
  - Self-management
  - Self-reflection
  - Results-oriented action
  - Role awareness
- Dealing with others:
  - Conflict resolution skills
  - Consulting and advisory skills
  - Collaborative skills
  - Communication skills
  - Stance/attitude
  - Ethical/moral stance/attitude

Aside from the basic competences, the Sifa competence profile (DGUV 2018) also features a list of subcompetences and a description of the specific form these competences take for safety professionals. Know-how, for example, refers not only to isolated technical expertise but also to the application of this expertise using specific methods as part of a systematic approach. That is because the complex challenges in today's world of work cannot be solved without effectively working together with partners inside and outside the company. Inside the company, safety professionals often have to work hard to persuade managers, sometimes coming up against strong resistance. That is why collaborative, communicative, and conflict resolution skills are so essential.

Furthermore, when implementing the training model, it became increasingly clear that consulting and advisory skills play a key role. During the first revision of the competence profile, therefore, advisory skills and the development of an advising strategy were given a higher priority.

The changing world of work thus also finds its reflection in the Sifa competence profile. Today's safety professionals embrace a comprehensive approach to occupational safety and health, they are connected to other professions and institutions, and they work to improve their company's structure and culture of prevention in a goal-oriented and systematic manner. Thus, for a Sifa to actually be able to perform their duties in the organization, they need to have both profound technical expertise and strong consulting competence. Without expert knowledge on safety and health, they will not be able to support their organization; without consulting competence, their technical expertise will fail to translate into the desired outcomes. As a consequence, there must be a good balance of technical expertise and consulting skills.

11.3.3 Suitable Structure and Culture of Prevention

Effective occupational safety and health measures with an emphasis on prevention must be aligned with appropriate operational procedures and structures, and they have to be understood as an integral part of a company's structure. In the training program, future safety professionals learn how to design and implement such organizational structures and how to identify the tasks they will have to address. In this respect, the Sifa training is based on the basic principles of the new ISO 45001 norm that specifies the requirements for an occupational health and safety management system (ISO 2018).

Ultimately, what matters in terms of occupational safety and health is not only installing a suitable management structure but also the culture of prevention prevailing in this organizational structure. How can the patterns based on values and norms and the basic assumptions of people within an organization be described? What are the starting points for improving a culture of prevention? The Sifa training program explores these questions, ideally enabling future safety professionals to get themselves and other people excited about embracing safety and health as core values in their decisions and actions.

11.4 FROM COMPETENCE PROFILE TO TRAINING DESIGN

11.4.1 Goals of Sifa Training

The goal of the Sifa training program is to enable future safety professionals to perform their tasks effectively. This raises the question of how the Sifa training program must be designed so that learners indeed acquire the target competences described in the competence profile. Four guidelines for teaching were developed to ensure the goals of the Sifa training program are accomplished. These guidelines guide the actions of those developing and implementing the training design.

- Design the Sifa training program around competence:

     In all of its components, the training design is guided by whether participants have acquired the necessary technical, methodological, social, and personal competences to be able to act as safety professionals in an organization in accordance with their role and the demands placed on them, and to do so successfully. From the outset, the goal is to have participants take an active, reflective, and responsible share in the expansion of their competences, for example, by fostering and utilizing realistic assessments of the relevant competence levels and their competence for self-organized learning.
- Enable self-organized learning:

     Wherever possible, participants develop their competences in a self-organized manner. "Participants actively acquire new competences, they already experience their increased competence during the training program, they apply the new competences in exemplary training situations, and they reflect on the viability of these competences in practice and their own contributions to this end" (DGUV 2011a).
- Adopt a real-world approach to training:

     "The structure of the training program is guided by mastering the complex tasks and problems encountered in practice, not by artificially isolated areas of theoretical expertise on occupational health and safety" (DGUV 2011a).
- Choose methods and media that support active, self-directed learning:

     Methods suitable for self-directed learning are primarily those methods that support learning with all the senses. Such methods include project work, simulation games, role play, case methods, and exploration, for example.

11.4.2 GENERAL FRAMEWORK AND SCOPE OF THE REDESIGNED TRAINING PROGRAM

In 1997, the Federal Ministry of Labor decided that the Sifa training program would take place as a blended learning activity at three different learning sites: in the classroom, in periods of self-directed study, and at participants' workplaces (BMA 1997). This setup also applies to the redesigned Sifa training program. The ministry's 1997 memorandum also included directions regarding the scope of the program.

However, since frequent absence from the workplace is a critical issue to many employers, a compromise had to be reached regarding the number of scheduled classroom days for the redesigned training program. Whereas the previous version of the program featured a total of 22 days of face-to-face training (five seminars) for the commercial sector and five days (two seminars) for the public sector in levels I and II, 20 days of face-to-face instruction (seven seminars) were agreed for the redesigned Sifa course.

The periods of self-directed study consist of six modules of roughly 280 hours; learning at the workplace is scheduled at four modules totaling some 280 hours as well. Overall, therefore, the Sifa training program consists of approximately 720 hours. At the statutory accidence insurance institutions, working professionals can complete these hours in a part-time program lasting about 1.5 years, whereas commercial providers sometimes also run the program as a full-time course of study.

11.4.3 SCENARIO-BASED APPROACH

As postulated in the guidelines on teaching, learners should be able to build on existing skills, and the program should be as real-world and action-oriented as possible. That is why learners, from the beginning to the end of the Sifa training program, consistently progress through a series of scenarios, so-called action situations, which mirror the real tasks of a safety professional working in an organization as closely as possible. Action situations thus provide the central theme throughout the program. For each action situation, a set of performance expectations was developed with an eye on the target competences in the competence profile. The performance expectations define the requirements that a safety professional must meet in order to be able to master the action situation successfully. Going from one learning area and action situation to the next, learners eventually acquire all target competences of the Sifa program (Figure 11.1).

11.5   SELF-ORGANIZED LEARNING ON THE "SIFA LERNWELT" PLATFORM AND AT THE WORKPLACE

11.5.1 LEARNING PLATFORM: SIFA LERNWELT

The training program takes place at three learning sites: learning in the seminar room, self-organized learning in the office or elsewhere, and learning at the workplace; all three sites are closely interconnected. The central element in this blended learning arrangement is a learning platform used in all three sites. The "Sifa Lernwelt" platform features a library offering Sifa knowledge modules and other media and materials to facilitate knowledge acquisition. In a virtual sample company, learners can practice the tasks of a safety professional, and the Sifa Lounge provides a forum for learners to communicate with each other or with their trainers.

Unlike many other learning platforms, which purely serve as document storage, "Sifa Lernwelt" is an interactive tool for learners to use over the entire course of their training. The materials and media needed for Sifa training are made available to all accredited training providers via the DGUV's Learning Content Management System (LCMS), meaning that all providers use the same materials.

11.5.2 LEARNING PATHS FOR PERIODS OF SELF-ORGANIZED LEARNING IN THE OFFICE

Understanding learning as an individual process means that learners themselves are responsible for this process. The second guideline on teaching ("enable self-organized learning") calls for keeping this in mind in the design of the training program.

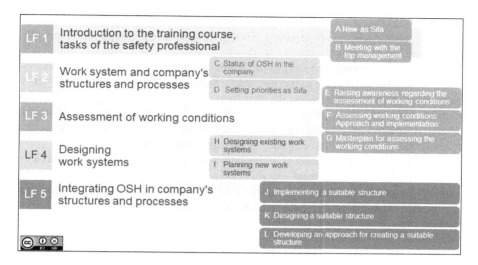

FIGURE 11.1   Learning fields (LF) and action situations in the Sifa training program.

This means the course must be designed in a way to actually enable learners to engage in self-organized and self-directed learning (see also Brendebach, Chapter 4 in this book). This applies to all learning sites, not just periods of self-organized learning in the office. But it is in the latter periods that the redesigned nature of the program becomes especially evident: The structured self-study units from the previous program are now redesigned as periods of self-organized learning. In the process of development, a balance had to be found between full self-directed learning on the one hand and efficient organization and learner support through pre-structuring and guidance on the other. The outcome was a concrete learning path guiding learners through the program and providing them with recommendations, especially during the periods of self-organized learning in the office. The training model provides for the following learning cycle: acquire – practice – apply – review.

11.5.3   ACQUIRE: SIFA KNOWLEDGE MODULES FOR IN-DEPTH TECHNICAL EXPERTISE

In-depth technical expertise is essential for any safety professional. The "Sifa Lernwelt" library contains about 100 Sifa knowledge modules for learners to acquire basic knowledge on the subject. In the near future, the knowledge modules are expected to be made available online to all interested learners, not just participants of the training program. In this way, the DGUV makes the concept of lifelong learning a reality, providing all stakeholders in occupational safety and health with a constantly updated overview of basic technical expertise.

11.5.4   PRACTICE: INSIDE A VIRTUAL COMPANY

It is important that learners practice concrete tasks. The practice situations were to be designed in a way that allows for in-depth exploration and – as in real-world situations working as a safety professional – in a way that helps participants under-stand the interdependencies involved. To this end, a virtual sample company was established.

The sample company is conceived as a company in the metal processing industry, which allows for simulating numerous work situations: production, logistics, office cleaning, group offices, garden maintenance, services performed at the client's premises, or road safety. Ultimately, work situations from a wide range of different areas may be connected to the sample company. Ideas for a company day care center or a neighboring hospital have already been suggested, for example.

Each of the work situations contains a complete field of activity, presented in an audiovisual format. Learners use this environment to become familiar with the company's concrete procedures, to assess working conditions, to share in the design of new work systems, and to help implement a suitable structure. In doing so, they practice the technical expertise they have acquired along with the corresponding methods.

Training providers may be interested in creating additional work situations. For this purpose, a manual has been prepared to illustrate how this is done. The plan is to make such work situations available in a central pool, thereby enabling all training providers to expand the range of work situations they can use.

11.5.5   APPLY AND REVIEW NEW COMPETENCES AT THE WORKPLACE

Learning in the redesigned Sifa training program takes place with a real-world orien-tation. This also means that participants start using their new expertise at their cur-rent employer; reflecting on this experience helps consolidate their newly acquired competences. This approach is intended to facilitate a successful and, above all, sustainable acquisition of competences.

However, the approach is not necessarily a linear one. When practicing or applying what they have learned, learners may notice they need to pick up addi-tional skills or abilities to be able to proceed effectively. Likewise, learners may notice during the phase of acquisition that they need to switch from acquisition to practice in order to be able to fully understand what they have learned in the first place and then continue with acquiring new knowledge. Reflection plays a critical role here: How well did I do when applying my new competences? What worked particularly well and why? What did not work so well and why? Which area of expertise and which skills and abilities do I need to work on more intensively?

11.5.6   TESTING TRAINING OUTCOMES THROUGH LEARNING ASSESSMENTS

Learners are required to take five learning assessments to demonstrate they have actually acquired the new competences. Like the program itself, the individual learning assessments are connected to each other. In assessment 1, learners evaluate a work situation in the virtual sample company, demonstrating they have acquired the necessary technical expertise to do so. In assessment 2, learners evaluate work-ing conditions at their own company. In other words, assessments 1 and 2 encompass the first steps of evaluating working conditions, a process that is continued in assessments 3 and 4.

Prior to learning assessment 5, learners explore procedures for (further) developing safety management at their respective companies. In assessment 5, they are asked to reflect on the point in the management project at which they did not make the desired progress. This challenge is the focus of assessment 5, which involves learners giving advice to each other.

As the pilot program has shown, learners embrace this advisory setup: they do want to help their classmates with their problems. Some even became somewhat oblivious of the fact that this was an examination. Nearly all participants said they were satisfied with the advice they received and claimed to have identified new ways of approaching the issue in their respective organizations.

The learning assessments are conceived as a cumulative evaluation. Learners have to pass all five assessments to earn the certificate at the end of the program. Each assessment highlights individual, independent aspects of the competences to be demonstrated; in sum, the assessments cover the full competence profile of a safety professional.

11.6   FROM TRAINING DESIGN TO THE COMPETENCE PROFILE OF THE LEARNING FACILITATOR

The redesigned training program, and especially the aforementioned guidelines on teaching, requires both trainers and learners to rethink the teaching-learning process. Self-organized and self-directed learning calls for a high degree of self-driven activity and motivation on the part of learners. Under this new paradigm, they must take ownership of their learning process and competence acquisition.

11.6.1   THE CHANGING ROLE OF TRAINERS: FROM INSTRUCTOR TO FACILITATOR

For trainers, this means a fundamental change in their role – a shift away from being "instructors" towards becoming "learning facilitators". They support learners' competence acquisition, they support and moderate the learning process, and they ensure a motivating learning environment.

The training design of the Sifa course must strike a fine balance between a prescribed curriculum and emphasizing a learner-centered approach to teaching. It reveals the ambivalence between a non-prescriptive design of the learning process that meets the needs of adult learners on the one hand and the need to achieve verifiable competence targets on the other.

To master this balancing act, the existing rigorous guidelines with their detailed specifications in terms of teaching were replaced with more flexible "learning arrangements". These require that learning facilitators receive appropriate training – not only regarding their new role but also regarding the use of the new media and learning sites. When developing the learning arrangements and the educational media, the perspective of the learning facilitators was consistently taken into account, and the competence profile for the learning facilitator was developed in parallel to that of the safety professional.

Because people acquire new knowledge, perspectives, and experiences against the background of the interpretative patterns adopted over the course of their biographies and in the context of their own learning projects, there is a strong need for trainers to create the conditions for self-organized learning and to enable processes of self-directed, independent knowledge acquisition.

(Arnold 2007, 73, author's translation)

Using activating methods in a suitable educational context and in response to the learners' specific needs and goals is a promising approach in this regard. However, it would be unrealistic to expect trainers to meet the demands of the learning facilitator role without providing them with appropriate training. According to Rolf Arnold, such a transformation of learning culture is only possible if the essential question of a trainer's professionalization has been answered: "How can prospective adult educators be encouraged to overcome the waiter's attitude of simply serving what's on the menu and move towards a professional attitude of assisted learning?" (ibid., 78, author's translation).

11.6.2 TRAINING PLAN TO ENSURE THE COMPETENCE PROFILE OF LEARNING FACILITATORS

Providers implementing the redesigned training program for safety professionals work with a wide range of learning facilitators with very different competences and backgrounds. They often hire experts for whom adult education is only a part of their activities. Then there are professional educators who may already have the relevant attitudes and skills but possibly lack in-depth knowledge of the subject. Just as prospective safety professionals should have a good balance of technical and advisory competence, a similar balance between technical and teaching competence is also desirable for learning facilitators. The competence profile for learning facilitators in the Sifa training program was developed accordingly (DGUV 2017).

This competence profile for learning facilitators served as the basis for developing a specific training plan. To qualify for this training, trainers must already have subject-related knowledge in the field of occupational safety and health, as well as basic knowledge in the field of learning facilitation.

The training program for Sifa learning facilitators consists of a basic module mandatory for all learning facilitators, which provides a basic understanding of the Sifa training program and its procedures. Learning facilitators get to know the structure of the training, they become familiar with the learning platform and the media modules, and they are introduced to the Sifa competence profile and the basic principles of competence measurement. The other modules address specific areas of activity.

In line with the concept of "lifelong learning", all education providers are called upon to recognize and implement the professionalization of their teaching staff as an ongoing responsibility. In addition to opportunities for self-reflection, collegial advice, and, if necessary, regular supervision, providers must offer continuous professional development and staff training opportunities to improve the quality of teaching in a long-term and sustainable manner. This is the only way for trainers to successfully develop the right internal "stance/attitude" and hence for the entire organization to move forward towards a sustainable and competence-oriented understanding of education – a transformation of educational culture that must also include the statutory accidence insurance institutions.

11.7 INITIAL EXPERIENCES FROM THE PILOT

In 2018, the revised concepts were put to practice for the first time: From March 2018 until April 2019, the training program was piloted at the DGUV's Institute for Work and Health (IAG) in two courses with 20 participants each.

The courses were evaluated systematically, involving a combination of written feedback from learners and learning facilitators, submitted via specifically created feedback forms, and feedback discussions in the seminar groups after each face-to-face period. The feedback discussions were led by external moderators who were not involved in the development and implementation of the Sifa training program. In addition, observers from the project group that developed the training program were present during the pilot and added their observations to the evaluation. Whereas the written feedback forms were only used by participants and learning facilitators during the first modules, the feedback discussions evolved into a central forum for learner feedback. As the results of the pilot have shown, the redesigned training program works and participants showed pronounced role awareness early on in the training and applied the comprehensive approach as a matter of course. As planned, future safety professionals were enabled to acquire the necessary competences. Furthermore, they engaged in intensive reflections on their role as supporters and consultants and recognized the key relevance of a suitable company structure when it comes to safety and health.

11.8 CONCLUSION AND OUTLOOK

The first set of accidence insurance institutions switched to the redesigned Sifa training program in the second half of 2019; in 2022, the transition is expected to be completed at all training providers.

The redesigned program thus took more than ten years to develop – much longer than originally anticipated. The development of media and materials turned out to be particularly time-consuming. One lesson learned by the project group was that the design of educational media has to be part of the process from the very beginning. Owing to the rapid advances in media technology, however, the prolonged development period also enabled the inclusion of new, state-of-the-art educational media. The highly realistic simulations of work situations, for example, provide learners with excellent opportunities to practice what they have learned in real-world scenarios.

11.8.1 CHANGES IN THE WORLD OF WORK AND CHANGES IN SIFA TRAINING

Work and society are in a process of constant change. This change also affects what is expected of safety professionals. The redesigned Sifa training program creates the necessary foundations to enable safety professionals to adapt their knowledge to changing needs and situations. As a result, they are also prepared to more easily acquire additional competences independently at a later point ("lifelong learning"). Transformation processes occurring inside the organization are another area where safety professionals can make active contributions. Likewise, safety professionals are capable of actively exploring and addressing health and safety issues brought on by digitalization, demographic change, new technologies, and the like, each in response to different needs and fields of activity.

However, the changes in the world of work also imply ongoing changes to the Sifa training program – changes that go beyond updating legal references and technical insights. Accordingly, the Sifa competence profile must be reviewed and adapted on a regular basis, and the Sifa training program must continue to evolve.

To achieve the maximum effect in terms of prevention, the program's contents and methods should be revisited by prevention specialists on a regular basis. The experiences and trends from the Sifa training program can be of interest with regard to regulation, much in the same way that regulation forms the basis for the Sifa training program. Likewise, engaging in international exchanges on the insights and experiences from safety professional training in other countries is another rewarding pursuit.

REFERENCES

Arnold, R. 2007. *Ich lerne, also bin ich. Eine systemisch-konstruktivistische Didaktik.* Heidelberg: Carl-Auer-Systeme-Verlag.

BMA – Federal Ministry for Employment. 1997. Schreiben des BMA an die Träger der gesetzlichen Unfallversicherung vom 29.12.1997 – III b 7-36042-5-. In *DGUV Report 2/2012: Training to be a Specialist in Health and Safety at Work, Optimization 2012*, eds. German Social Accident Insurance (DGUV) and Federal Institute for Occupational Safety and Health (BAuA), 2nd ed., 281–283. Meckenheim: DCM.

DGUV – German Social Accident Insurance. 2009. *Accident Prevention Regulation Occupational Physicians and OSH Professionals: DGUV Regulation 2.* Berlin: German Social Accident Insurance. https://www.dguv.de/medien/inhalt/praevention/vorschriften_regeln/regulation_2_en.pdf (accessed December 14, 2019).

DGUV – German Social Accident Insurance. 2011. *Accident Prevention Regulation Occupational Physicians and OSH Professionals. DGUV Regulation 2, Agreed Sample Text.* https://www.dguv.de/medien/inhalt/praevention/vorschriften_regeln/dguv-vorschrift_2/muster_vorschr_2.pdf (accessed September 25, 2019).

DGUV – German Social Accident Insurance. 2011a. *Training Model for the Training of Safety Professionals.* https://www.dguv.de/medien/sifa-online/documents/ausbildungsmodell_komplett.pdf (accessed September 25, 2019).

DGUV – German Social Accident Insurance. 2012. *Training to be a Specialist in Health and Safety at Work, Optimization 2012, DGUV Report 2/2012*, eds. German Social Accident Insurance (DGUV) and Federal Institute for Occupational Safety and Health (BAuA), 2nd ed. Meckenheim: DCM.

DGUV – German Social Accident Insurance. 2017. *Competence Profile of Learning Facilitators in Sifa Training*, authors C. Eickholt, W. Hamacher, G. Riering, and A. Wegener, eds. German Social Accident Insurance (DGUV) and Federal Institute for Occupational Safety and Health (BAuA). https://www.dguv.de/medien/sifa-online/dokumente/kompetenzprofil_lernbegleitung_2205.pdf (accessed September 25, 2019).

DGUV – German Social Accident Insurance. 2018. *Competence Profile of the Safety Professionals*, authors C. Eickholt, W. Hamacher, G. Riering, A. Wegener, M. Schröder, and R. Reitz. Berlin: DGUV. https://www.dguv.de/medien/sifa-online/dokumente/kompetenzprofil_2018-07-10_v2.pdf (accessed December 14, 2019).

Hamacher, W., N. Lenartz, S. Riebe, R. Trimpop, and K. Höhn. 2013. *Prävention wirksam gestalten – Erkenntnisse aus der Sifa-Langzeitstudie. DGUV Report 3/2013.* Rheinbreitbach: Medienhaus Plump.

Heyse, V., and J. Erpenbeck. 2009. *Kompetenztraining. Informations- und Trainingsprogramme*, 2nd ed. Stuttgart: Schäffer-Poeschel.

ISO – International Organization for Standardization. 2018. *ISO 45001: 2018. Occupational Health and Safety Management Systems -- Requirements with Guidance for Use.* Geneva, Switzerland: ISO.

Weinert, F. 2001. Concept of competence: A conceptual clarification. In *Defining and Selecting Key-Competences*, eds. D. Rychen, and L. Salganik, 45–65. Seattle, WA: Hogrefe & Huber.

12 School Heads as Change Agents
Salutogenic Management for Better Schools

Peter Paulus and Heinz Hundeloh

CONTENTS

12.1 INTRODUCTION

School heads play an important role in school development. This is a complex task and requires specific competences. This also includes competence in school health management. The responsibility for school health management has grown in importance in recent years because of the increasingly critical health situation of students and schoolteachers, as well as the heads of school, especially in highly developed Western countries.

Health in school is more than another aspect of health promotion. It is closely related to the core mission of schools, namely teaching and learning. There is a causal relationship between a healthy environment and the quality of the work teachers do and of students' success (Suhrcke and de Paz Nieves 2011; Dadaczynski 2012). Health in school is determined by factors related to school structures and the processes of school life. It is therefore crucial that the school in its educational *and* organizational dimensions becomes a salutogenic setting – more resource than risk oriented – that supports education and learning. This chapter identifies the necessary competences of school heads in developing such a safe and healthy setting, in which safety aspects are integrated as part of a broader concept of school health. In the following, the holistic concept of the "good healthy school" serves as a conceptual framework to describe the role and responsibility of school heads.

12.2 SCHOOL HEADS: NEW CHALLENGES FOR THEIR ROLE AND RESPONSIBILITIES

Neoliberal forms of the welfare state organization, now permeating society, are increasingly impacting educational systems in Europe. Politically initiated educational reform efforts are gaining ground almost worldwide (for the global educational reform movement, see Sahlberg 2015). These reforms share the attempt to increase the efficiency of the educational system and to modernize it in the face of social, economic, and technological transformations, some of which are conflict laden. These transformations have significantly changed the activities and contexts of school management.

As key players in the design of the school, school heads face new challenges in this process and are under pressure to meet them. New output-oriented control models have led to an intensification of the professional demands placed on them, an expansion of their responsibilities, and to a monitoring by state authorities of their performance and that of the schools through external evaluations. On the one hand, their powers and areas of responsibility have expanded as a result; on the other hand, they are given more scope for action and can make more autonomous decisions in various areas. Increasingly, however, school heads are also held responsible for the performance of this system within the framework of a decentralized yet autonomous management of the individual school. In addition to their work as managers of the school and its teachers, school heads are therefore now also responsible for the quality of the school and its further development. For example, in Germany, they must struggle with scarce resources and new challenges, including the recent implementation of all-day schools, the introduction of inclusive pedagogy (Sowada and Terhart 2015), and the influence of digital media. These rapidly spreading developments are beginning to massively change school life and teaching. Moreover, school heads are also required to position themselves together with their school within a framework of "managed competition" in the new educational governance structures (Abs et al. 2015) and to have their work reviewed by external evaluations (Schratz et al. 2016).

This means that the work of school heads has become more diversified, compared with the traditional demands placed on them. They now include new duties and responsibilities, including human resource management and development, teaching-related leadership, and organizational management and development (Huber and Rolff 2010; for "system leadership", see Huber 2014). Fulfilling these responsibilities will require a new bundle of competences, which will put the measures of school management in a new light.

School heads are therefore one of the most important drivers of the development of teaching and schools, making them "change agents" (Sowada and Terhart 2015;

Ärlestig et al. 2016). In this role, school heads see themselves as designers, facilitators, and developers in a cooperatively organized dialogical process of "transformational leadership", thereby combining the management of the school (their planning, organization, leadership, and management) with the creation and development of a culture of cooperation and self-responsibility based on the relationships among its members in a jointly developed vision of a good school.

However, school heads can only convincingly fulfill their diverse duties and responsibilities as managers if their own health, above all their mental health, and the associated work-related personal and organizational resources they need as health-maintaining factors are not endangered. The following section describes the health situation of students, teachers, and school heads first. The reason for this is the interrelation between the new requirements and the burden on the health of school leaders and the health situation of students and teachers. The description is mainly based on data from Germany. Afterwards, the task of school health management and the necessary competences of school heads are described.

12.3 THE HEALTH SITUATION OF STUDENTS, TEACHERS, AND SCHOOL HEADS IN GERMANY

The number of worrying national and international findings on the safety and health of students, teachers, and school heads is increasing. The morbidity spectrum in highly developed Western countries has shifted from acute diseases to a general increase in mental health and behavioral problems. In the area of student safety at school, the national social accident insurance reported 877,628 accidents at general schools with 8.4 million students in total. That means that one in ten students is involved in an accident (DGUV 2017).

12.3.1 SCHOOL STUDENTS

Overweight and obesity alongside associated chronic diseases such as diabetes or cardiovascular diseases have increased significantly (Kolip et al. 1995; Dadaczynski et al. 2018; Dadaczynski 2018a), as well as headache, abdominal and back pain (Krause et al. 2017), and conspicuous eating habits or individual core symptoms of an eating disorder (Pickhardt et al. 2019). Bullying at schools has also become an everyday phenomenon. Every sixth 15-year-old in Germany is bullied at school (OECD 2017; UNICEF 2018). Psychological abnormalities (emotional problems, problems with peers, behavioral problems, hyperactivity) in children and adolescents are also on the rise (Klipker et al. 2018).

These significantly negative quantitative values are of particular importance in the school context since they are closely related to the demands school places on the cognitive-perceptual and social-emotional-motivational disposition spectrum of students. Such variances are all the more disturbing if they occur when children begin school because they can then have a negative influence not only on the initial phase of schooling but on the further course of the child's education.

However, at a level of 20% to over 40%, schools are increasingly perceived as stressful by the students (Beisenkamp et al. 2011; IFT-Nord 2017; Berngruber et al.

2018). Circumstances experienced as stressful include the teacher-student relationship (whereby stress, depression, and burnout symptoms by teachers have consequences for the students), engagement in school councils, and school grounds and its service facilities (e.g., schoolyard, school canteen). It is also important to what extent a school can convey perspectives and meaningful experiences and create a positive social atmosphere in the school and classroom (Hascher 2004; Bilz 2008; Rathmann et al. 2018).

12.3.2 Teachers

Teaching is a profession with a particularly high potential for excessive workloads and stress. As a result, mental health and its risks are prominently represented in the empirical findings. Numerous papers have already been published demonstrating these findings (Schaarschmidt 2005; Krause et al. 2010; Lehr 2014; Scheuch et al. 2015; Wesselborg 2015; Rothland and Klusmann 2016). In rankings of the factors leading to stress, school students experienced as difficult are at the top (Wesselborg 2015), typically followed by lack of class discipline (Ksienzyk and Schaarschmidt 2005), high teaching load and time pressure (Mußmann et al. 2017), high classroom noise levels (Nübling et al. 2012), and, increasingly often, dealing with difficult parents.

These studies therefore testify a high risk for teachers of mental and psychosomatic stress, which can lead to illness and even accidents. Stress symptoms that are more common among teachers compared with other workers include fatigue/exhaustion, nervousness/irritability, headaches, sleep disorders, hearing deterioration, and tinnitus. Compared with other professions, there is also an increased incidence of cognitive stress symptoms, such as problems with attention span, making decisions, memory, and thinking clearly (Nübling et al. 2008).

Studies investigating determinants of the professional demands placed on teachers demonstrate the occupational hazards teachers face, which are localized in schools as social organizations with their management structures and processes and in the specific working conditions and support resources available to them (e.g., collegial coaching, see Meißner et al. 2019). A comprehensive overview of risk factors as a result of a narrative review of existing national and international studies on teacher health describes those factors (Dadaczynski and Nieskens 2014). They form the basis for the development of survey methods with which teachers can systematically assess mental health risks they experience at work. As reported, the 2013 version of the European Labor Protection Act now in force in the European Union stipulates that mental health risks in schools must now also be recorded and that school management is responsible for their implementation and also for the measures to be taken after evaluation (Table 12.1).

Determinants can also be found in the person of the teacher (Badura 2004; Krause and Dorsemagen 2014), for example, in their professional expectations of self-efficacy (Rothland and Klusmann 2016) or in their knowledge of classroom management methods (König and Rothland 2016). Together with the coping styles and strategies described above, the complex interaction of individual and organizational determinants leads to physical and/or mental health situations for teachers.

TABLE 12.1

Dimensions and Criteria of a Self-Assessment Procedure for Recording the Mental Health Risks Teachers Face

Work content/task	Work organization	Social relationships	Work environment	New forms of work
Quantitative demands	School management and leadership	Interaction with parents	Teaching and learning materials	Compatibility of family and work
Qualitative demands	Breaks and free hours	Interaction with students	Equipment	
Inclusion and heterogeneity	(Collegial) decision-making	Interaction with colleagues	Rooms	
Freedom of action	School and lesson organization	Recognition, appreciation, and feedback	School environment	
Emotional demands	School culture and shared values			
Professional development				

12.3.3 School Heads

School heads themselves are also stressed by their professional activities, and these can have consequences for their health. A representative study shows first of all that the job satisfaction of school heads – regardless of gender, age, or type of school – is very high at around 95% (forsa 2018). However, other more extensive studies of school heads in various federal states in Germany (n = 4,326; Dadaczynski 2014; Dadaczynski and Paulus 2016) show that the wellbeing of school heads has declined in recent years. Approximately 20% of all respondents show a lower level of wellbeing compared with 2009, and about 12% have a very low level of wellbeing. However, the level of wellbeing demonstrated by these studies is still higher than teachers' self-assessment. School heads often suffer from emotional and motivational exhaustion. When asked about the causes of health problems, school heads report that it is above all the workload and the regulations of the school supervisory authorities and those of external school inspections that have a negative impact, in contrast to the demands of working with colleagues, which is perceived as having little impact (Dadaczynski 2014; Paulus et al. 2017).

12.4 HEALTH AND EDUCATION

As discussed above, the health problems of school students, teachers, and school heads are serious. However, this is not the only reason why these problems are significant. A further reason is that, depending on the type of health impairment, these problems can massively impact teaching and learning in schools or, in a positive case, also be conducive to and promote educational and work processes.

In this case health, and mental health in particular, is an important resource and driver of education; that is, it is an essential prerequisite and foundation for the pedagogical work of teachers and school management in schools, but also, closely related, for the learning of school students (Paulus 2009, 2010; Retelsdorf et al. 2010; Shen et al. 2015). Healthy school students learn better, and healthy teachers teach better.

Moreover, in general, teachers showing higher quality work and results (Belz 2008; Frenzel et al. 2008) are able to better motivate their students (Knauder 2005) and better support them (Klusmann et al. 2006). In other words, the importance of health at school goes far beyond the conventional understanding of health education. In this concept, health is no longer the outcome or output of health education teaching and methodology; it is an input, a throughput factor of education in general, and not only a driver of health education (Paulus 2010). This relationship has been shown by Kevin Dadaczynski (2012; see also for an earlier version Suhrcke and de Paz Nieves 2011) in a review of longitudinal studies (see Figure 12.1).

Other findings point to the mediating function of mental disorders in educational processes and outcomes. For example, stigmatization and discrimination as a result of overweight or obesity in childhood and adolescence encourage the internalizing and externalizing of psychological problems, which in turn act as mediators jointly responsible for educational inequalities (Crosnoe 2007; Dadaczynski et al. 2018).

The importance of teachers in this context is shown by a number of further studies. If teachers feel comfortable at school and are satisfied with the prevailing general psychosocial atmosphere at their school, the quality of their instruction increases, which in turn promotes student performance because they feel supported by the teacher in their learning behavior (Belz 2008; Frenzel et al. 2008; Klusmann et al.

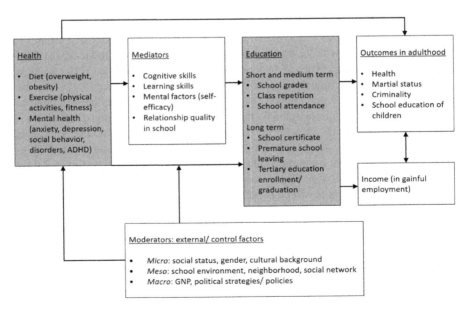

FIGURE 12.1   The relationship between health and education (Suhrcke and de Paz Nieves 2011; Dadaczynski 2012).

2016; Klusmann and Waschke 2018). Teachers who can relate to their students, who radiate emotional warmth, who respond empathetically to their students and treat them with trust, who are enthusiastic and embody commitment, promote the wellbeing, the joy of learning, and the educational success of their students (Knauder 2005; Peperkorn 2019). These teachers are, moreover, those whom Uwe Schaarschmidt described in his studies as the "healthy ones" and who are rated by the school students as the most popular.

The results of the research reported here on the influence of health on educational processes and outcomes make one thing very clear: schools must deal more strongly than before with the fact that health concerns are part of everyday school life and have an impact on their core mission of teaching and learning. The importance and explosive nature of the link between health and education is also illustrated by the effects of the introduction of inclusivity in schools. Today, schools are confronted with a wide diversity of health problems, making it a challenge to continue to ensure that education for all is possible – as stated already by Comenius in his "Didacta Magna" published in 1657.

12.5 HEALTH MANAGEMENT AS A RESPONSIBILITY OF SCHOOL HEADS

The argumentation in this article illustrates how central the topic of health is for schools. It is a cross-cutting issue that is closely linked to issues of school development and quality assurance. This position is also represented by the German Conference of the Ministers of Education and Cultural Affairs in its recommendation on health promotion and prevention in schools (KMK 2012) when it states: "The school heads have a central function and responsibility in the implementation of health management and health promotion within the framework of school personnel and organizational development" (ibid., 2).

These responsibilities come into focus when it becomes clear what health management is all about. It is about the systematic and sustainable development of school framework conditions, school structures, and processes that focus on the health-appropriate or health-compatible design of teaching, learning, and organization. It is also about the ability of all groups of people involved in the school to act in a health-promoting manner, while in turn health management measures contribute to maintaining and promoting the health of all school employees (Dadaczynski and Paulus 2011). The double meaning of the term "health management" should be considered here. The predominant use of the term health management means "managing health", but it can also be understood to mean "managing in a healthy way" and thus focuses on the way in which health management is conducted. This is referred to again in the last section of this chapter.

The management of health, as well as all of the management responsibilities of the school head, is thus geared to the fulfillment of the school's educational goals. "With health create a good school" is a maxim that describes the ultimate goal of health management. It means identifying appropriate health-related interventions that will also give the actions of the school head a health-conscious quality and that will lead to processes and structures of a school organization designed to be health-effective. It is therefore not a question of short-term measures or of individual actions such as appointing a hazardous materials officer or developing and practicing a fire escape plan at school. Instead, health management should be understood as "school development". It is a cross-sectional responsibility that includes lesson development, human resource development, and organizational development.

Many school heads consider such a concept of health management as a crucial factor for their schools. In a survey conducted by Dadaczynski (2014), school heads from the Federal State of North Rhine-Westphalia say that they have a high interest in the health of their school students and teachers. At the same time, it is clear that school heads do not believe they have sufficient resources to deal adequately with these problems. For example, they report that there is often a lack of regular training on health-related topics. With regard to their competences, there are therefore clearly uncertainties among school heads as to how measures of school-based prevention and health promotion are to be implemented.

Personal attitudes towards school-based prevention and health promotion, perceived moral responsibility, and beliefs about one's own competences and experiences of professional self-efficacy explain more than one-third of the differences in the intention of school heads to support measures of school-based prevention and health promotion and prevention in the future. When the intentions reported by the school heads and their perceived personal competence and professional self-efficacy are included, about 30% of the variance of their implementation of health management measures can also be explained (see Figure 12.2).

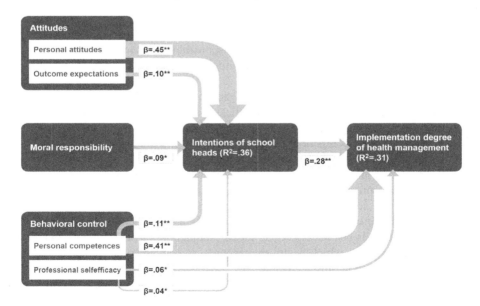

FIGURE 12.2 School management conditions of school health management (Dadaczynski 2014, 21). Arrow width is a function of the strength of each relationship: the thicker the arrow is, the stronger the relationship. Level of significance, $p \leq .001$; $p \leq 0.05$.

12.5.1 HEALTH MANAGEMENT TARGET GROUPS

The target groups of health management in schools are the teachers, the non-teaching staff, the school students, but also the school heads themselves (who are often not addressed). Regarding the self-management of school heads, the general conditions of attitude, behavioral control, and moral responsibility mentioned above are also significant.

12.5.2 HEALTH MANAGEMENT MEASURES

In addition to target-group orientation, health management also requires behavioral measures that focus on the development of personal resources (e.g., competence training) and behavior-related measures that include the health-stabilizing design of school work and the learning environment (e.g., ergonomic school furniture, but also a safety and health promoting school culture). These measures are implemented in the areas of (1) occupational safety and health: measures to assess risks and minimize or eliminate them; (2) prevention and health promotion: school development or project management measures to strengthen and promote resources; and (3) crisis and emergency management: crisis prevention measures, crisis intervention, and post-crisis care.

In the tradition of health promotion, these different measures are characterized by (1) a holistic approach (e.g., reduction of stressful working conditions, creation of safety and health-promoting working conditions, reduction of stressful and health-risk behavior, development of individual health potentials); (2) participation by all school members; (3) integration of education and health in the perspective of a "good healthy school"; (4) systematic project management; (5) orientation to the quality framework of the respective federal state/province; and (6) gender mainstreaming.

12.6 CREATING A HEALTHY SCHOOL ENVIRONMENT

The previous discussion of empirical findings emphasizes the importance of health for educational processes in schools and shows how necessary and meaningful it is to develop schools in their organizational structure and in their educational processes and structures into a place of salutogenesis. A salutogenically designed school involves more than seeing health as an outcome of efforts undertaken in a school, as it has been conceived in traditional school education (Wulfhorst 2001; Stroß 2012). Nor is it a question of the extent to which the school, as a social organization and as an architecturally designed environment, is a determinant of the state of health of school members, that is, as a factor making them sick or healthy. Instead, it is about the extent to which the design of schools in accordance with health criteria (or the neglect or disregard of them) influences school education. From a salutogenic perspective, it is not only a question of designing lessons so that healthy teaching and learning can become a real possibility; it is also about designing a school with a safety and health-promoting culture, a healthy psychosocial atmosphere, professionalization opportunities for teachers in the field of health, healthy leadership, and school management, and the physical and technical infrastructure to promote safety and health.

Health is thus understood more broadly as the foundation of school, as embedded in the structures and processes of school work and as an integral part of a school seen as a good school. A commonly used model of school quality in Germany (according to Ditton 2002) is shown in Figure 12.3.

The health quality of a school thus extends what is commonly understood as a "good school", in school effectiveness research, for example, to its health quality aspects. This fusion is expressed in the concept of the "good healthy school" of the German Social Accident Insurance (DGUV 2013). A good healthy school is a school "which in its development has clearly committed itself to the quality dimensions of the good school and which specifically uses health interventions in the realization of its educational mission. The aim is to increase the quality of education in schools in a sustainable and effective way" (Paulus 2009, 2010; Hundeloh 2012). Such an approach can then be understood as organizational health management, as Rosenbusch (2013) conceived it, in that health acts as a crucial condition, a driver, and an intermediate goal in school development (Huber 2013).

From the perspective of a good healthy school, the problems of school-based health promotion are twofold: "Which health interventions are appropriate to our school?" and "How can they be implemented?" Apart from possible interventions that have been discussed above, the question of implementation is equally important. Health promotion as a driver of school quality and thus of educational quality is the duty and responsibility of the school head and the school management team. In order to actually be able to have an impact in all areas of school quality, school health promotion must be integrated into all quality dimensions as a criterion and made measurable by means of indicators. This is how school health promotion can become part of school development. In organizational terms, it should be integrated into the school's development team, whose task it is to ensure the quality of the school and to develop it further.

An example will illustrate this procedure. The starting point is the quality dimension No. 6 "teaching and learning" of the "Quality framework of the School" (Ditton 2002), with generally accepted criteria which are underpinned by indicators that enable pedagogical quality to be measured. For example (1) school support (e.g., teachers strive to make their lessons comprehensible and meaningful for their students and to strengthen their competences (e.g., self-efficacy, self-esteem); (2) learning in and outside of the classroom (e.g., extra-curricular learning venues with a health dimension are also used to ensure that safety and health is taught and practiced in an interdisciplinary manner – e.g., sports clubs, museums); (3) the school's own internal curriculum (e.g., health is integrated into the school's own curriculum; and (4) the spatio-temporal design for learning (e.g., attention is paid to good lighting conditions, good room atmosphere [ventilation], and protection against pollutants, noise reduction, cleanliness and hygiene [e.g., toilet facilities] [Paulus 2009, 21 et seq.]). These criteria, together with their indicators, describe measures that from a health science perspective are key to a school integrating safety and health aspects into its quality assurance plan and developing into a good healthy school.

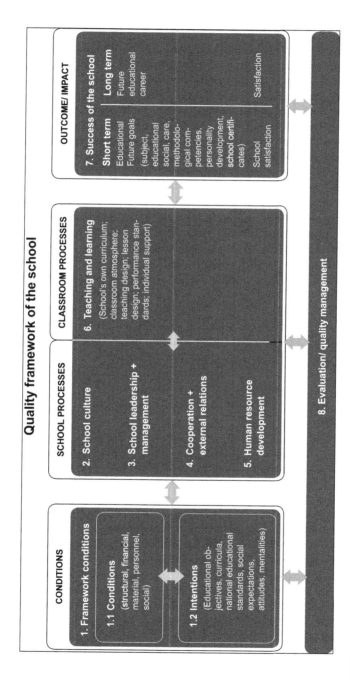

FIGURE 12.3 Model of the eight dimensions of school quality (Ditton 2002).

12.7   THE ROLE OF THE SCHOOL MANAGEMENT

The prominent position of the school head in the new management models was described above. It is their responsibility, together with the school management team, to disseminate the idea of a good healthy school to the teachers and staff in order to gain acceptance and support for the project of developing a good school with health. This line of thinking might at first meet with little appreciation, since traditional school health promotion measures are often seen by the teaching staff as an "add-on" activity to other tasks imposed on teachers and the school that must be dealt with and, in the best case scenario, mastered. Communicating the fact that health interventions are meant as a support for teachers, as an "add-in" to help them fulfill their core mission of good teaching and good schooling, is an important task for school management.

This communication process must be designed in such a way that teachers can experience how the application of health science knowledge can advance school development projects. One promising way of designing communication is to "harness" health management to a leading idea, which then determines the design of the processes and structures needed to implement health in the school. Now everything depends on the *how*, on "managing in a healthy way". How can different aspects of health, the *what*, be introduced and integrated into the quality development of the school? A leading idea that could do this is what Aaron Antonovsky calls a "sense of coherence" (1997). Generally speaking, the sense of coherence describes a feeling of fundamental trust in the world. It is composed of three elements: (1) a "sense of comprehensibility" in which demands from the inner and outer world of experience in the course of life are experienced as structured, predictable, and explainable; (2) a "sense of manageability" in which resources are available that are necessary to meet these demands; and (3) a "sense of meaningfulness" that allows these demands to be seen as challenges deserving investment and commitment.

Applied to the school, Antonovsky's concept means that the school head promotes internal and external communication, information, and predictability, thus conveying a sense of comprehensibility; that the school head helps his/her staff to meet the requirements by providing them with sufficient resources and thus creating a sense of control over the work processes; and that the school head has finally developed values aimed at the best possible integration of the individual goals of the teaching and non-teaching staff and the common goals of the school, thus conveying a sense of meaningfulness. The school head can achieve this through direct or indirect actions, communication, and work organization.

Such management actions can be described as salutogenic. They comprise leading a school both with regard to management and administration of the entire organization and its subareas, as well as its processes and structures with regard to the management of the individual employees with the explicit inclusion of health-relevant findings (Hundeloh 2012). It is characterized by the aspects of communication and work organization, in which the dichotomy of behavior and relations are seen as the constituent elements of holistic, integrated strategies of school-based prevention and health promotion.

Since a sense of coherence is an essential personal resource for mental health, such leadership not only promotes the mental health of school members, but also, when applied to the health-conscious design of the school, promotes the integration of education and health on the path to becoming a good healthy school.

This concept of salutogenic leadership fits into the comprehensive concept of transformational leadership already mentioned at the beginning of this article (see also Mields and Birner, Chapter 3; Brendebach, Chapter 4; Schöbel, Chapter 5 in this book). It is based on a sustainable improvement of the problem-solving, learning, and performance capacity of the school with the participation of all its members. Open communication and cooperation enables them to participate, supports responsible joint problem-solving processes and goal-oriented cooperation, releases potential and allows innovative ideas to emerge, gives sustainable impetus to the development of the school, and is supported by the school community itself. This is confirmed by research findings (Achermann Fawcett et al. 2018) that increasingly indicate that these leadership concepts, initiated and implemented by the school management or the school management team, have a positive correlation with job satisfaction (Griffith 2004), commitment (Geijsel et al. 2003), and the wellbeing of the teachers as well as negatively with burnout (Harazd et al. 2009; Harazd and van Ophuysen 2011; Gerick 2014). Salutogenic management offers school heads a promising way to transform their schools into good healthy schools.

12.8   SUMMARY AND OUTLOOK

With the changes in the demands and challenges placed upon schools today, the roles and functions of school heads and school management teams have changed significantly (Schratz et al. 2016). They have become more complex and demanding. The burdens associated with leadership have led to psychological stress not only for school heads but also among teachers and pupils. Health management has therefore become a school management responsibility in its own right. The concept of a "good healthy school" with the core thesis that health – and in particular mental health – functions as a foundation, resource, and a driver of school education for all those involved in the school system, forms the basis for a specific orientation of health management in schools. The concept of "salutogenic leadership", based on Aaron Antonovsky's sense of coherence model, shows how health management can become "healthy management" for the whole school and how health can effectively promote education in school development.

REFERENCES

Abs, H. J. T., M. Brüsemeister, M. Schemmann, and J. Wissinger, eds. 2015. *Governance im Bildungswesen. Analysen zur Mehrebenenperspektive, Steuerung und Koordination.* Wiesbaden: Springer.

Achermann Fawcett, E., R. Keller, and P. Gabola. 2018. *Bedeutung der Gesundheit von Schulleitenden und Lehrpersonen für die Gesundheit und den Bildungserfolg von Schülerinnen und Schülern. Wissenschaftliche Grundlage für das Argumentarium "Gesundheit stärkt Bildung".* Zürich and Lausanne: Allianz Betriebliche Gesundheitsförderung in der Schule.

Antonovsky, A. 1997. *Health, Stress, and Coping: New Perspectives on Mental and Physical Well-Being*. San Francisco: Jossey- Bass.

Ärlestig, H., C. Day, and O. Johannsson, eds. 2016. *A Decade of Research on School Principals: Cases from 24 Countries*. Wiesbaden: Springer.

Badura, B. 2004 *Auf dem Weg zu gesunden Schulen: Was Schulen dabei von Unternehmen lernen können*. Gutachten für die Landesregierung NRW. Bielefeld: Universität Bielefeld.

Beisenkamp, A., K. Müthing, S. Hallmann, and Ch. Klöckner. 2011. *Elefanten-Kindergesundheitsstudie 2011. Große Ohren für kleine Leute*. Herten: Institut für Sozialforschung.

Belz, C. 2008. Bewältigungsstrategien von Belastungen, Stressoren und Konflikten. In *Sportpädagogik im Spannungsfeld gesellschaftlicher Erwartungen, wissenschaftlicher Ansprüche und empirischer Befunde*, eds. V. Oesterhelt, J. Hofmann, M. Schimanski, M. Scholz, and H. Altenberger, 289–292. Hamburg: Czwalina.

Berngruber, A., N. Gaupp, and A. N. Langmeyer. 2018. Lebenswelten von Kindern und Jugendlichen. In *Datenreport 2018. Ein Sozialbericht für die Bundesrepublik Deutschland*, ed. Statistisches Bundesamt and Wissenschaftszentrum Berlin für Sozialforschung, 86–92. Bonn: Bundeszentrale für politische Bildung.

Bilz, L. 2008. *Schule und psychische Gesundheit. Risikobedingungen für emotionale Auffälligkeiten von Schülerinnen und Schülern*. Wiesbaden: VS Verlag für Sozialwissenschaften.

Crosnoe, R. 2007. Gender, obesity, and education. *Sociology of Education* 80(3), 241–260.

Dadaczynski, K., and P. Paulus. 2011. Psychische Gesundheit aus Sicht von Schulleitungen: Erste Ergebnisse einer internationalen Onlinestudie für Deutschland. *Psychologie in Erziehung und Unterricht* 58(4), 306–318.

Dadaczynski, K. 2012. Die Rolle der Schulleitung in der guten gesunden Schule. In *Handbuch Lehrergesundheit: Impulse für die Entwicklung guter gesunder Schulen*, eds. DAK and UK NRW, 2nd ed., 197–228. Köln: Carl Link Verlag.

Dadaczynski, K. 2014. *Schulleitung und Gesundheit: Zur gesundheitlichen Lage von Schulleitungen in NRW*. Prävention in NRW, Band 53. Düsseldorf: Unfallkasse Nordrhein-Westfalen.

Dadaczynski, K., and B. Nieskens. 2014. *Psychische Gefährdungsbeurteilung an Schulen in NRW. Beurteilung der Prüflisten zu vorwiegend psychischen Belastungen im Lehrerberuf*. Lüneburg: Leuphana Universität Lüneburg. Zentrum für angewandte Gesundheitswissenschaften.

Dadaczynski, K., and P. Paulus. 2016. Wohlbefinden von Schulleitungen in Deutschland: Ausprägungen und Zusammenhänge mit Arbeit und Gesundheit. *Prävention und Gesundheitsförderung* 11(3), 171–176.

Dadaczynski, K., O. Backhaus, and P. Paulus. 2018. Der Einfluss des Gewichtsstatus von Kindern und Jugendlichen auf Bildungsoutcomes. In *Übergewichtsprävention im Kindes- und Jugendalter. Grundlagen, Strategien und Interventionskonzepte in Lebenswelten*, eds. K. Dadaczynski, E. Quilling, and U. Walter, 95–108. Göttingen: Hogrefe.

Dadaczynski, K., E. Quilling, and U. Walter, eds. 2018a. *Übergewichtsprävention im Kindes- und Jugendalter Grundlagen, Strategien und Interventionskonzepte in Lebenswelten*. Göttingen: Hogrefe.

DGUV – German Statutory Accident Insurance. 2013. *Strategic Concept: "Using Health to Develop Good Schools*, ed. Specialist Division Educational Institutions. Berlin: German Social Accident Insurance.

DGUV – German Statutory Accident Insurance. 2017. Facts and figures. https://www.dgu v.de/en/facts-figures/index.jsp (accessed December 15, 2019).

Ditton, H. 2002. Evaluation und Qualitätssicherung. In *Handbuch Bildungsforschung*, ed. R. Tippelt, 775–790. Opladen: Leske and Budrich.

forsa – Politik- und Sozialforschung. 2018. *Die Schule aus Sicht der Schulleiterinnen und Schulleiter – Berufszufriedenheit von Schulleitungen.* Berlin: Forsa.

Frenzel, A. C., T. Götz, and R. Pekrun. 2008. Ursachen und Wirkungen von Lehreremotionen: Ein Modell zur reziproken Beeinflussung von Lehrkräften und Klassenmerkmalen. In *Lehrerexpertise: Analyse und Bedeutung unterrichtlichen Handelns*, eds. M. Gläser-Zikuda, and J. Seifried, 187–209. Münster: Waxmann.

Geijsel, F., P. Sleegers, K. Leithwood, and D. Jantzi. 2003. Transformational leadership effects on teacher's commitment and effort toward school reform. *Journal of Educational Administration* 41(3), 228–256.

Gerick, J. 2014. *Führung und Gesundheit in der Organisation Schule: Zur Wahrnehmung transformationaler Führung und die Bedeutung für die Lehrergesundheit als Schulqualitätsmerkmal.* Münster: Waxmann.

Griffith, J. 2004. Relation of principal transformational leadership to school staff job satisfaction, staff turnover, and school performance. *Journal of Educational Administration* 42(3), 333–356.

Harazd, B., M. Gieske, and H.-G. Rolff. 2009. *Gesundheitsmanagement in der Schule.* Köln: Link Luchterhand.

Harazd, B., and S. van Ophuysen. 2011. Transformationale Führung in Schulen. Der Einsatz des "Multifactor Leadership Questionnaire" (MLQ 5 x Short). *Journal of Educational Research* 3(1), 141–167.

Hascher, T. 2004. *Wohlbefinden in der Schule.* Münster: Waxmann.

Huber, S. G., and H.-G. Rolff. 2010. Delegation und system leadership. In *Führung, Steuerung, Management*, ed. H.-G. Rolff, 43–54. Seelze: Kallmeyer.

Huber, S. G., ed. 2013. *Jahrbuch Schulleitung 2013. Befunde und Impulse zu den Handlungsfeldern des Schulmanagements.* Köln: Wolters Kluwer.

Huber, S. G. 2014. School leadership and leadership development – Adjusting leadership theories and development programs to values and the core purpose of school. *Journal of Educational Administration* 42, 669–684.

Hundeloh, H. 2012. *Gesundheitsmanagement an Schulen. Prävention und Gesundheitsförderung als Aufgaben der Schulleitung.* Weinheim: Beltz.

IFT-Nord Institute for Therapy and Health Research GmbH. 2017. *Präventionsradar. Kinder- und Jugendgesundheit in Schulen.* Kiel: IFT-Nord.

Klipker, K., F. Baumgarten, K. Göbel, T. Lampert, and H. Hölling. 2018. Psychische Auffälligkeiten bei Kindern und Jugendlichen in Deutschland – Querschnittergebnisse aus KiGGS Welle 2 und Trends. *Journal of Health Monitoring* 3(3), 37–45.

Klusmann, U., M. Kunter, U. Trautwein, and J. Baumert. 2006. Lehrerbelastung und Unterrichtsqualität aus der Perspektive von Lehrenden und Lernenden. *Zeitschrift für Pädagogische Psychologie* 26(4), 161–173.

Klusmann, U., D. Richter, and O. Lüdtke. 2016. Teacher's exhaustion is negatively related to student`s achievements: Evidence from a large scale assessment study. *Journal of Educational Psychology* 108(8), 1193–1203.

Klusmann, U., and N. Waschke. 2018. *Gesundheit und Wohlbefinden im Lehrerberuf.* Göttingen: Hogrefe.

KMK – Conference of Ministers of Education and Cultural Affairs. 2012. *Recommendation for Health Promotion and Prevention at School: Resolution of the Standing Conference of the Ministers of Education and Cultural Affairs of 15.11.2012.* Berlin: KMK. https ://www.kmk.org/kmk/information-in-english/standing-conference.html (accessed December 15, 2019).

Knauder, H. 2005. *Burn-out im Lehrerberuf. Verlorene Hoffnung und wiedergewonnener Mut.* Graz: LeyKam.

Kolip, P., K. Hurrelmann, and P. E. Schnabel. 1995. *Jugend und Gesundheit – Interventionsfelder und Präventionsbereiche.* Weinheim: Juventa.

König, J., and M. Rothland. 2016. Klassenführungswissen als Ressource der Burnout-Prävention? Zum Nutzen von pädagogisch-psychologischem Wissen im Lehrerberuf. *Unterrichtswissenschaft* 44(4), 425–441.

Krause, A., L. Meder, A. Philip, and H. Schüpach. 2010. Gesundheit, Arbeitssituation und Leistungsfähigkeit der Lehrkräfte. In Bildungsförderung durch Gesundheit. Bestandsaufnahme und Perspektiven für eine gute gesunde Schule, ed. P. Paulus, 57–85. Weinheim: Juventa.

Krause, A., and C. Dorsemagen. 2014. Belastung und Beanspruchung im Lehrerberuf. Arbeitsplatz- und bedingungsbezogene Forschung. In *Handbuch der Forschung zum Lehrerberuf*, eds. E. Terhart, H. Bennewitz, and M. Rothland, 2nd ed., 987–1013. Münster: Waxmann Verlag.

Krause, L., H. Neuhauser, H. Hölling, and U. Ellert. 2017. Kopf-, Bauch- und Rückenschmerzen bei Kindern und Jugendlichen in Deutschland – Aktuelle Prävalenzen und zeitliche Trends: Ergebnisse der KiGGS-Studie: Erste Folgebefragung (KiGGS Welle 1). *Monatsschrift für Kinderheilkunde* 165(5), 416–426.

Ksienzyk, B., and U. Schaarschmidt. 2005. Beanspruchung und schulische Bedingungen. In *Halbtagsjobber? Psychische Gesundheit im Lehrerberuf – Analyse eines veränderungsbedürftigen Zustandes*, ed. U. Schaarschmidt, 2nd ed., 72–87. Weinheim, Basel: Beltz.

Lehr, D. 2014. Belastung und Beanspruchung im Lehrerberuf – Gesundheitliche Situation und Evidenz für Risikofaktoren. In *Handbuch der Forschung zum Lehrerberuf*, eds. E. Terhart, H. Bennewitz, and M. Rothland, 2nd ed., 947–967. Münster: Waxmann Verlag.

Meißner, S., I. Semper, S. Roth, and N. Berkemeyer. 2019. Gesunde Lehrkräfte durch kollegiale Fallberatung? Ergebnisse einer qualitativen Evaluationsstudie im Rahmen Des Projekts "Gesunde Lehrkräfte durch Gemeinschaft". *Prävention und Gesundheitsförderung* 14(1). doi:10.1007/s11553-018-0684-8.

Mußmann, F., Th. Hardwig, and M. Riethmüller. 2017. *Niedersächsische Arbeitsbelastungsstudie 2016. Lehrkräfte an öffentlichen Schulen.* Göttingen: Kooperationsstelle Hochschulen und Gewerkschaften.

Nübling, M., M. Wirtz, R. Neuner, and A. Krause. 2008. Ermittlung psychischer Belastungen bei Lehrkräften. Entwicklung eines Instruments für die Vollerhebung in Baden-Württemberg. *Zentralblatt Arbeitsmedizin* 58(10), 312–313.

Nübling, M., M. Vomstein, A. Haug, T. Nübling, U. Stößel, H. M. Hasselhorn, and A. Krause. 2012. *Personenbezogene Gefährdungsbeurteilung an öffentlichen Schulen in Baden-Württemberg. Erhebung psychosozialer Faktoren bei der Arbeit.* Freiburg: Freiburger Forschungsstelle Arbeits- und Sozialmedizin.

OECD – Organisation for Economic Co-operation and Development. 2017. *PISA 2015 Results (Volume III): Students' well-being.* Paris: OECD.

Paulus, P. 2009. *Anschub.de – ein Programm zur Förderung der guten gesunden Schule.* Münster: Waxmann.

Paulus, P., ed. 2010. *Bildungsförderung durch Gesundheit. Bestandsaufnahme und Perspektiven für eine gute gesunde Schule.* Weinheim München: Juventa.

Paulus, P., D. Horstmann, C. Baydar, and K. Dadaczynski. 2017. *Abschlussbericht zur OnlineBefragung "Mehr Zeit für gute Schule".* Lüneburg: Leuphana Universität. Zentrum für Angewandte Gesundheitswissenschaften.

Peperkorn, M. 2019. *Lehrkräftegesundheit im Kontext schulischer Inklusion.* PhD, Leuphana University Lüneburg.

Pickhardt, M., L. Adametz, F. Richter, B. Strauß, and U. Berger. 2019. Deutschsprachige Präventionsprogramme für Essstörungen – Ein systematisches Review. *Psychotherapie, Psychosomatik, Medizinische Psychologie* 69(01), 10–19.

Rathmann, K., M. Herke, K. Heilmann, J. M. Kinnunen, A. Rimpelä, K. Hurrelmann, and M. Richter. 2018. Perceived school climate, academic well-being and school-aged children's self-rated health: A mediator analysis. *European Journal of Public Health* 28(6), 1012–1018.

Retelsdorf, J., R. Butler, L. Streblow, and U. Schiefele. 2010. Teachers' goal orientations for teaching: Associations with instructional practices, interest in teaching, and burnout. *Learning and Instruction* 20(1), 30–46.

Rosenbusch, H. S. 2013. Organisationspädagogische Führungsprinzipien. In *Handbuch Führungskräfteentwicklung. Grundlagen und Handreichungen zur Qualifizierung und Personalentwicklung im Schulsystem*, ed. S. G. Huber, 96–103. Köln: Carl Link / Wolters Kluwer.

Rothland, M., and U. Klusmann. 2016. Belastung und Beanspruchung im Lehrerberuf. In *Beruf Lehrer/Lehrerin: Ein Studienbuch*, ed. M. Rothland, 351–369. Stuttgart: UTB.

Sahlberg, P. 2015. *Finnish Lessons 2.0: What Can the World Learn from Educational Change in Finland?* 2nd ed. New York: Teachers College Press.

Schaarschmidt, U., ed. 2005. *Halbtagsjobber? Psychische Gesundheit Im Lehrerberuf – Analyse eines veränderungsbedürftigen Zustandes*, 2nd ed. Weinheim: Beltz.

Scheuch, K., E. Haufe, and R. Seibt. 2015. Lehrergesundheit. *Deutsches Ärzteblatt* 111(20), 347–356.

Schratz, M., C. Wiesner, D. Kemethofer, A. C. George, E. Rauscher, and S. Krenn. 2016. Schulleitung im Wandel: Anforderungen an eine ergebnisorientierte Führungskultur. In *Nationaler Bildungsbericht Österreich 2015*, eds. M. Bruneforth, F. Eder, K. Krainer, C. Schreiner, A. Seel, and C. Spiel, Volume 2: Fokussierte Analysen bildungspolitischer Schwerpunktthemen, 221–262. Graz: Leykam.

Shen, B., N. McCaughtry, J. Martin, A. Garn, N. Kulik, and M. Fahlman. 2015. The relationship between teacher burnout and student motivation. *British Journal of Educational Psychology* 85(4), 519–532.

Sowada, M. G., and E. Terhart. 2015. Schulinspektion und Unterrichtsentwicklung. In *Handbuch Unterrichtsentwicklung*, ed. H.-G. Rolff, 195–208. Weinheim / Basel: Beltz.

Stroß, A. M. 2012. Gesundheitserziehung und –bildung als Handlungsfelder einer reflexiven Gesundheitspädagogik. Geschichte, Gegenwart, Perspektiven. In *Handbuch Bildungs- und Erziehungssoziologie*, eds. U. Bauer, U. H. Bittlingmayer, and A. Scherr, 741–761. Wiesbaden: Springer.

Suhrcke, M., and C. de Paz Nieves. 2011. *The Impact of Health and Health Behaviours on Educational Outcomes in High Income Countries: A Review of the Evidence.* Copenhagen: WHO Regional Office for Europe.

UNICEF – United Nations International Children's Emergency Fund. 2018. *An Everyday Lesson: #END Violence in Schools.* https://www.unicef.org/publications/index_103 153.html (accessed December 15, 2019).

Wesselborg, B. 2015. *Lehrergesundheit: Eine empirische Studie zu Anforderungen und Ressourcen im Lehrerberuf aus verschiedenen Perspektiven.* PhD. University Tübingen. Baltmannsweiler: Schneider-Verlag.

Wulfhorst, B. 2001. *Theorie der Gesundheitspädagogik. Legitimation, Aufgabe und Funktion von Gesundheitserziehung.* Weinheim: Juventa.

Author Index

Subjet Index

Printed in the United States
by Baker & Taylor Publisher Services